IntechOpen

Automation and Control - Theories and Applications

Edited by Elmer P. Dadios

Published in London, United Kingdom

Supporting open minds since 2005

Automation and Control - Theories and Applications
http://dx.doi.org/10.5772/intechopen.95125
Edited by Elmer P. Dadios

Contributors
Barbara Arbanas, Stjepan Bogdan, Frano Petric, Antun Ivanović, Marko Car, Ana Batinović, Lovro Marković, Ivo Vatavuk, Ivan Hrabar, Marsela Polić, Murat Aydın, Igor N. Sinitsyn, İhsan Ömür Bucak, Elmer P. Dadios, Sandy Lauguico, Ryan Rhay Vicerra, Argel Bandala, Ronnie Concepcion II, Edwin Sybingco

First published in London, United Kingdom, 2022 by IntechOpen
IntechOpen is the global imprint of INTECHOPEN LIMITED, registered in England and Wales, registration number: 11086078, 5 Princes Gate Court, London, SW7 2QJ, United Kingdom
Printed in Croatia

British Library Cataloguing-in-Publication Data
A catalogue record for this book is available from the British Library

Additional hard and PDF copies can be obtained from orders@intechopen.com

Automation and Control - Theories and Applications
Edited by Elmer P. Dadios
p. cm.
Print ISBN 978-1-83969-173-7
Online ISBN 978-1-83969-174-4
eBook (PDF) ISBN 978-1-83969-211-6

For EU product safety concerns:
IN TECH d.o.o., Prolaz Marije Krucifikse Kozulić 3, 51000 Rijeka, Croatia, info@intechopen.com or visit our website at intechopen.com.

Meet the editor

Dr. Elmer P. Dadios obtained his Ph.D. from Loughborough University, United Kingdom. He is a Distinguished Professor and University Fellow of De La Salle University, Manila, Philippines. He is a Professorial Chairholder of the Thomas J. Lee Chair in Manufacturing Engineering Management and Victor T. Lu Chair in Production Management. He is the founder and chairman of the board of Neuronmek Inc. and Intelligent Systems Innovation Corporation. He is a top 100 scientist in the Philippines, according to the AD Scientific Index 2022. He was also listed as a top 100 scientist in 2019 by Asian Scientist Magazine. Dr. Dadios is the recipient of the 2018 Philippine Association for the Advancement of Science and Technology D. M. Consunji Award for Engineering Research; Lifetime Achievement Award from the National Research Council of the Philippines (NRCP); and the Department of Science and Technology (DOST) 50 Men and Women of Science and Technology. Currently, Dr. Dadios serves as editor of the Journal of Advanced Computational Intelligence and Intelligent Informatics (JACIII) and the Journal of AI, Computer Science, and Robotics Technology. He is the editor-in-chief of the Journal of Computational Intelligence in Engineering Applications (JCIEA). He was the chair of the IEEE Philippines Section from 2010 to 2012, and the founder and chair of the IEEE Computational Intelligence Society - Philippines Chapter and the IEEE Robotics and Automation Society (RAS) - Philippines Chapter. He was also an EXCOM member of the IEEE Asia-Pacific Region from 2012 to 2020. He was the chair of the IEEE Asia-Pacific Region Awards and Recognition Committee from 2016 to 2020.

Contents

Preface

Automation and control are the powers or abilities to make something or someone performs a task in a complex and changing environment. Advances in automation and control today cover many areas of technology where human input is minimized. These areas include not only industry and manufacturing but also agriculture management and business-specific types of automation such as IT, policy, and regulatory automations. Automation and control are also used for personal applications such as games. This book discusses numerous types and applications of automation and control.

Chapter 1 deals with automation and control for adaptive management of urban agriculture using computational intelligence. The insufficiency in food production is a global challenge that needs to be addressed, as emphasized by the United Nations Sustainable Development Goal 12 (SDG-12). One of the most feasible solutions to the global food shortage is the establishment of farms in urban areas. Urban agriculture (UA) is a concept that has been put forth to promote the planting and cultivating of crops within cities. This chapter presents an adaptive management system (AMS) that operates UA autonomously to provide an artificial environment suitable to grow and produce cultivars effectively.

Chapter 2 presents a review of building information modeling (BIM)–based automated code compliance checking (ACCC). A high level of interoperability and cooperation is essential to enhance expertise and automation in the construction industry. The required data must be represented correctly according to the types and characteristics of buildings. Manually conducted building regulation checking, a traditional method, is a repetitive, time-consuming, and error-prone process for architects, engineers, and public authorities. The BIM's effective automated code compliance checking is considered a promising domain in Architecture, Engineering, and Construction (AEC). In this chapter, the study fields are limited to architectural design, automation, building codes, engineering, environmental science, construction industry, and construction building technology. The results presented assert the importance of automated code compliance checking.

Chapter 3 presents a control algorithm using a heuristically generated metric approach to the solution of a chase problem (a pursuer and a prey). This is a cat-and-mouse game involving the cat and mouse finding each other, making smart moves, and making appropriate decisions. A fuzzy logic system controller is developed to obtain the distance and find a route between the cat and the mouse. A problem of this kind is of interest to the military community and video game developers.

Chapter 4 presents the development of an aerial-ground robotic team for a search-and-rescue operation. This team has been entered in the 2019 European Robotics League (ERL) Emergency Robots competition and the 2020 Mohamed Bin Zayed International Robotics Challenge (MBZIRC). Search-and-rescue problems for collaborative multi-robot systems have been an interesting research topic for several decades. The attractiveness of the domain stems from the variety of problems it incorporates, including mapping and situational awareness, monitoring and surveillance, establishing communication networks, and cooperative decision-making.

Finally, Chapter 5 presents the theory of stochastic control with unsolved derivatives. Different types of stochastic differential systems with unsolved derivatives (SDS USD) arise in problems of analytical modeling and estimation for stochastic control systems when it is possible to neglect higher-order time derivatives. The chapter shows methodological and algorithmic support for analytical modeling, filtering, and extrapolation for SDS USD. The methodology is based on the reduction of SDS USD to SDS by means of linear and nonlinear regression models.

Elmer P. Dadios
Department of Manufacturing Engineering and Management,
De La Salle University,
Manila, Philippines

Chapter 1

Automation and Control for Adaptive Management System of Urban Agriculture Using Computational Intelligence

Elmer P. Dadios, Ryan Rhay Vicerra, Sandy Lauguico, Argel Bandala, Ronnie Concepcion II and Edwin Sybingco

Abstract

It has been predicted by the United Nations that the world population will increase to 9.8 billion in 2050. This causes agricultural development areas to be transformed into urban areas. This urbanization and increase in population density cause food insecurity. Urban agriculture using precision farming becomes a feasible solution to meet the growing demand for food and space. An adaptive management system (AMS) is necessary for such farm to provide an artificial environment suitable to produce cultivars effectively. This research proposes the development of a computational intelligence-based urban farm automation and control system utilizing machine learning and fuzzy logic system models. A quality assessment is employed for adjusting the environmental parameters with respect to the cultivars' requirements. The system is composed of sensors for data acquisition and actuators for model-dictated responses to stimuli. Data logging was done wirelessly through a router that would collect and monitor data through a cloud-based dashboard. The model intended for training from the acquired data undergo statistical comparative analysis and least computational cost analysis to optimize the performance. The system performance was evaluated by monitoring the conditions of the sensors and actuators. Experiment results showed that the proposed system is accurate, robust, and reliable.

Keywords: urban agriculture, precision farming, adaptive control, automation, aquaponics, computational intelligence

1. Introduction

Increasing population density reduces land availability and quality [1, 2]. There is evidence that areas having higher population densities are correlated with having smaller farm sizes [3]. The correlation is highly apparent in urban areas. Urbanization has been rapid in the recent decades with the transformation or rural regions into urban areas [4]. Thus, as the population density increases, rural areas, which where agricultural land were originally based, are being rapidly developed into cities for accommodating the increasing demand of spaces for shelter and industrial infrastructures. It has been gradually noticeable that food production had not met the rural areas' expectation on its crucial role in different sectors [5], further proving that urbanization may impose challenges such as food insecurity [6].

Insufficiency in food production is a global challenge that needs to be addressed as emphasized by United Nation Sustainable Development Goal (SDG 12) with responsible consumption and production [7]. One of the most feasible solutions is the establishment of farms in urban areas (UA) to contribute for food security. The idea promotes the planting and cultivating of crops within cities [8, 9]. Moreover, it also involves complex systems that consider indoor food production inside factories with an artificial environment suitable for cultivation [10], which applies the discipline of a controlled environment agriculture (CEA) [11, 12].

The limited availability of space in urban is addressed by one of the common forms: the aquaponics (AP) system. Such a system is considered to be an emerging technique for combining intensive production with waste recycling and water conservation [13]. Common AP systems do not control their environment [14]. However, one usual challenge in AP is the management of nutrients in the water being shared by the crops and fish [15, 16]. This leads to the concept of using technological control and automation of the environmental parameters affecting growth and cultivation implementing the concept of CEA in AP systems. It is proven that since AP has a hydroponic component that does not require soil for cultivation, the use of CEA can optimize production and energy conservation [17].

Controlled environment agriculture is an intensive method for managing plant growth and development through taking advantage of technological advancements and innovations in horticulture [18]. Another issue to consider is that even though AP addresses conservation, CEA consumes a lot of energy for operations due to the use of innumerable devices from sensors to controllers [19]. Efficiency in farm performances is quantified from sustainable intensification defined to be the maximum ability of the system to produce [20]. Sustainable intensification models were proven to increase production and upgraded profits per unit of energy invested while maintaining the same consumption of energy [21]. There are numerous responses relating to sustainable intensification, involving the application of innovative technology to enhance control over factors such as nutrient use efficiency to reduce attribute-derived environmental risk [22]. Automating systems in the farming community are commonly operated with an expert system (ES) that is a computer program designed to emulate the logic and reasoning of a human expert through if-then rules as a tool for decision-making support [23]. ES-based automation is bound to a static configuration set by the programmer, resulting in a fixed control that does not respond on the real-time necessity of the system. The integration of urban agriculture principles and intelligent controller and automation may be beneficial down to the community level [24–31].

The specifc objectives of this study are as follows: (1) to implement wireless sensor nodes for irrigation control, nutrient mixture automation, adaptive temperature maintenance, and lighting systems between the hydroponic chambers and the pond for aquaculture; (2) to develop a smart control and automation on actuators based on the collected data from the sensors; (3) to wirelessly send the data acquired from the sensors to a common router node for cloud-based monitoring; (4) to develop a computational intelligence-based model in evaluating the performance of the smart automation system with respect to crop productions; and (5) to evaluate the developed model by determining the exhibited accuracy and sensitivity.

2. Developments of urban agriculture with its control and automation technologies

2.1 Urban agriculture perspective

Land resources for agricultural utilization are rapidly decreasing as they are developed and transformed into cities for accommodating the increasing demand of food due to drastic population growth [28]. Urbanization has both become a solution and a problem as it addresses land space issues for residential and industrial purposes, while causing lack of available land area for agriculture.

A study [30] on the success of urban farming concluded that city-adjusted farms in comparison with their counterpart are better in terms of three parameters: economic farm situations, positive farm prospects, and farm succession development. This results to foreseeing that urban agriculture (UA)-based businesses have small probabilities of decline and closure. The effectivity of UA in food production and business profitability has been evident. However, there are still questions on UA's capability to contribute in securing food demand. To further improve city farming performances, most of its advances are credited to the innovative technology for UA (ITUA), defined to be the integration of control and automation technological advancement for optimizing food production in open or closed systems [31]. Among different forms of ITUA, Aquaponics systems (AP) earned the most attention of researchers. Even though this is the case, the economic sustainability and feasibility of such systems remain an open research area and still require further extensive studies.

Utilizing control and automation technology and innovation in UA was proven to be beneficial in expanding access to food and agriculture [32]. Aside from automated food production, ITUA has been relevant in treating waste. Hydroponics systems (HD), which falls under ITUAs, was proven to enable and control decentralization of wastewater treatment, which in turn could provide nutrients for crops being cultivated in HDs using technology-based efficient removal of unnecessary nutrients (i.e., nitrogen and phosphorus) to sustain crop growth [33].

A new discipline identified as biosystems engineering (BE) is determined to be a major necessity to deal with ITUA as bioproducts and bioenergy will be produced through series of structures, operations, machines, converters that are well systematized on which most of the applications are biological in nature [34]. This concept applies with the feasibility of ITUA for food production and environmental sustainability; thus, application of technology in farming is enabled through BE.

Farming in urban contexts by default is not that sustainable for the environment because it is dependent on high-energy consumption and intensive capital needs particularly in controlled environment agriculture (CEA). The need for CEA relies on its advantage for being capable of optimized year-round production, higher yields, and improved water usage efficiency [35–43]. This, hence, focuses on production efficiency while neglecting environmental sustainability, which contradicts the feasibility of UA. Synthesizing the presented technologies, CEA should be operated through BE making it an ITUA-founded farming, which could result to improving energy consumption and maximizing financial capital while maintaining high-production performance despite CEA's heavy operational requirements.

2.2 Intelligent controller and automation applications for urban agriculture

The emerging advancements of CEA allowed to solve agricultural concerns ranging from climate change to food insecurity. One environmental parameter that remarkably affects growth of crops is humidity. Controlling such parameter is a necessity for overall productivity, sustainability, and energy efficiency in a CEA. A liquid desiccant system integrated with arrays of triple-bore hydrophobic hollow fiber membranes was developed to control humidity levels for maintaining an optimal environment suitable for plant cultivation [44]. Temperature is another noteworthy attribute that could influence plant growth. An fuzzy logic-based cooling system for tomato cultivation was developed in a soil-based close system to vary temperature of the environment based on growth stage and the time of the day for increasing crop productivity [45].

Artificial lighting systems have become a research focus of many studies to optimize and alternatively replace sunlight's contribution for photosynthesis in a close environment. One research aimed on exploring the consequences of a multichromatic light-emitting diode (LED) spectrum in a controlled environment chambers with regard to nutritive primary metabolites in green and red leaves of lettuce being cultivated [46]. The study concluded with identifying that metabolic plasticity of cultivars determines lettuce crops' sensitivity to lighting spectrum. Reduction of power consumption and attainment of optimal plant growth were put into consideration [47] in an automated indoor farming that utilizes far-red LED treatment. The technology produces variable lighting intensity through a micro-controller, solid state relay, and dimmable LED light for controlling flowering process and stem elongation.

Nutrient assessment up to production quality analysis is usually done in a CEA. A demonstration of altering macro-cation proportions in the nutrient solution (K/Ca/Mg proportions) was done to prove that it is possible to increase or enhance the concentration of the respective macro-minerals and key phytochemicals in lettuce crops and reduce anti-nutritional components such as nitrate regardless of crop genotype [48]. This was implemented through targeted modulation of cationic proportions in the nutrient solution, especially through the application of proportionally elevated magnesium. The development can deal with the demand for crops needing to have high nutritional value and enhanced bioactive content. Image analysis technique was utilized for measuring plant growth properties that are commonly grown in a CEA *via* a smartphone integrated to a local desktop [49]. The application contributes in monitoring and assessing quality of cultivated plants in a challenging controlled environment.

Water irrigation from recycling and reusing mixed mackerel and brown seaweed wastewater for cultivating lettuce crops was implemented in a hydroponic environment. Samples were measured from installed pH probes to determine high chlorophyll and carotenoid content and high antioxidant activity from lettuce to determine the effects of wastewater, therefore, ensuring crop quality and maximizing water resources [50]. A study on vision-based lettuce phenotype model using fuzzy logic controller integrated with fertigation system showed excellent nutrient efficacy and lower chemical wastewater emissions compared with manual fertigation [51].

Farming automation is the language used in technology-based urban agriculture whether in an open or a closed system, especially in a data-driven era. Agricultural economists are challenged with handling and analyzing big data that can determine specific actions or logical responses from the information obtained [52].

Recent research focuses on developing CEA that are fully automated. A study implemented a prediction model and was imparted for irrigation scheduling and automation to manage water usage for optimizing water resources through adjusting water content to the actual volume explored by the crops' roots [53]. The method involved estimating the root depth attained from digital photographs of the vegetation cover to analyze gains and losses of water to determine soil water status.

Another study focused on farm management that uses a multi-level automation for information system [54]. It was done by implementing three automation levels that improve farm management information systems (FMIS) *via* provision of solutions relating to the acquisition of fragmented-missing data and time-consuming data entry. It has contributed to effective financial analyses and assessment, task formulation, and profitability analyses. A fully automated hydroponics was also setup with the use of multiple sensors and microcontrollers. Android and iOS devices were also used to remotely monitor information from the sensors and provide analyses [55]. This significantly contributes to farming in remote places, which could be a basis for future research on places where agriculture is not usually set.

Monitoring automatically through wireless network communication is also becoming relevant in CEA. Internet of things (IoT) have made tremendous breakthroughs in farm automation. IoT provides the possibility of connecting all things to the Internet for various advantages such as remote monitoring and control, large data storage, and information accessibility. The use of IoT in agriculture is becoming more relevant. The performance of an indoor micro-climate horticulture farming was developed with the use of IoT for gathering data from sensors and for acquiring weather information from a meteorological agency for automating environmental factors in the farm [56]. Integrating IoT to unmanned aerial vehicles (UAV) is also used in open system or traditional agriculture to transform it to precision agriculture (PA) [57]. A study proved this technique to further improve crop yield and quality, reduce cost, and mitigate ecological footprint for traditional farming [58]. Data from the agricultural industry are contributing enormously in problem-solving as IoT opened ways for easy access of these. However, agricultural data can be messy, which could provide uncertain data quality resulting to inaccurate analyses. Preserving data in a secure storage was developed in helping farmers [59].

Aquaponics (AP) is the integration of hydroponic-based vegetable crop cultivation with an aquaculture unit for an innovative smart and sustainable production system, which plays a crucial role in the future of environmental and socioeconomic sustainability in smart cities [60]. The emerging AP systems have the potential to achieve high success rate. However, intensive monitoring, control, and management are essential to properly provide a conducive environment for all cultivars grown in both the hydroponics and the aquaculture systems [61]. Challenges in AP systems are difficult to address as a major factor involves the recirculating water from the marine system that is used for irrigation and gives nutrients to the crops planted in the hydroponics chambers, which in turn feeds back the water again to the aquaculture unit with different substances and nutrient concentration. An example of this phenomena is the fish wastewater, provided through recirculating aquaculture system (RAS), may cause to contain high amounts of microorganism that can compete with plants for oxygen [62], therefore not sufficiently providing the nutrients needed by the crops.

Performance assessment is a relevant AP research concept. A study [63] conducted focused on assessing how the three different AP systems carry out with its operation for small-scale production. The three configurations experimented with were Nutrient Film-Technique (NFT), floating raft, and vertical felt living wall system. Statistical analyses with SPSS 24 statistical package were performed. A comparative analysis was also done through one-way ANOVA and Shapiro–Wilk test assessed the normality of the data. Results showed that among the three systems, the NFT outperformed the rest in terms of crop production and water consumption. With regard to fish production, no significant differences in performance were observed. Through statistical analyses, results showed that nitrogen transformation, which includes water nitrogen retention, and nitrous oxide emission, is affected by the plant-fish (P/F) biomass ratio [64].

Since AP falls under the classification of CEA, automation plays a crucial role on operating such systems reliably. A recent study [65] aimed on including an innovative and sustainable AP system solution, a modular solution for an adaptable and a scalable local condition, and an optimal way of reusing water resources, and Supervisory Control and Data Acquisition (SCADA) and Manufacturing execution systems (MESs) were the techniques that operated the configuration. Through these, the collection of software and hardware components enabled the management to automate fish and crop production. An indoor farming configured with an automated AP system was designed and implemented [66] that can monitor and control the system through a mobile phone which resulted in highly successful vegetable yield. Further integrating the BE concept, an automated solar-powered AP system was developed [67]. Water quality, greenhouse environmental conditions, solar energy conversion status, and cooling and heating parameters were controlled and monitored through NI LabVIEW that was successful in considering the environmental impact of the setup while providing optimal yield.

The majority of automation systems in agriculture rely on expert system and static programming for control. Those systems are limited to the fixed standards of environmental parameters needed in the ecosystem. Applying adjustment in accordance to the real-time needs of the cultivars is not addressed. This may result to not catering the immediate necessity of produce, which yields to production and operation inefficiency. Nonetheless, automation is still an integral part of an IUAs, especially to CEAs. A quantitative research based on case studies and desk researchers analyzed existing data to present the need of a multisystem and multifaceted approach [67] to address the problem with regard to farmers not realizing that their decisions and actions toward agriculture are causing a reduction on economic efficiency, making automation play an important role for farming.

Optimizing automation relies on adaptivity. An adaptive management has long been known to ecology and conservation. Classical methods in biological conservation do not usually consider uncertainties in the state of a system and the model describing its dynamics [68]. The study included a solution for adaptive management of ecological systems. It is a significant strategy in addressing complex issues in natural resource management, which corresponds to decision making under uncertainty and uncertainty reduction through learning from arrival of new data. A sample application of the study involved fisheries' population mass is used to adjust harvest decisions.

Monitoring automation can also apply adaptive management system (AMS). A smart farm that applies remote monitoring adapts or adjusts to what wireless communication technology to be used between LoRaWAN and IEEE802.11 ac depending on their respective advantages [69, 70]. A classification application used in an adaptive farm topology develops a Naïve Bayes model for accurately identifying on which to allocate agricultural fields into different farm types [71].

Agriculture automation is an emerging concept in the industry. Together with the use of machine visions (MV) and its subsets: machine (ML) and deep learning (DL), has shown potential in solving different challenges in agriculture [72]. A lot of difficulties arise in the field such as crop pests, crop diseases, lack of irrigation control, weed, water, and storage mismanagement, and plant misclassification. Due to these, expert system-based control and automation are becoming less appropriate for addressing agricultural complications as this is limited to thinking processes. This gave birth to the use of an intelligent machine powered by the discipline of artificial intelligence (AI). AI is a field of computer science and engineering that ventures to reversing the human brain and is capable of maximizing rate of success for solving such problems by providing analytical decisions to determine the most suitable environment for sustaining plant growth based on learning from past experiences [73].

Advancing biosystems engineering with AI can enable agriculture machinery engines for a more comprehensive automation especially in CEA. Various machine learning techniques have been engaged in precision agriculture for both supervised learning and unsupervised learning in assessing plant health status and condition and invasive plant species recognition through the use of spectral signatures and optical features [74]. According to the study, there are three major contributions to agriculture that data analytics-based ML can provide. These are as follows:

1. Crop status for optimal production can be done by fusing information from spatial, spectral, and time series of crop parameters for detecting trends related to the condition of crops.

2. Hardware sensors and actuators can also be improved by making them compact and embeddable in field-deployable devices. Integrating with the discipline makes use of the devices for acquiring big data for real-time analysis, which can allow event-based decision algorithm to automatically respond for managing crop conditions on a real-time basis.

3. Accurate and reliable models can be trained from the data gathered for assessing and predicting future potential states within the field. Applications from weather projections to soil maps can be made possible for determining crop suitability in the examined fields.

With the presented contributions, studies have shown specifically the important use of AI in different agricultural applications. A smart pesticide sprayer was designed and developed with the use of AI and MV in a traditional agriculture. The sprayer was attached to an all-terrain vehicle autonomously driving itself with the aid of a global positioning system. The MV systems function as the target detector. Once a target is locked in with the system processing, the algorithm, which uses YOLOv3 and convolutional neural network (CNN), instructs the end effector to spray [75]. The study is fully automated and does not require a farmer to manually spray pesticides to each of the crops with defects.

A deep reinforcement learning, an AI subset, was used merging with IoT to enable a smart farm to make immediate decisions such as determining the amount of water that needs to be irrigated for enhancing the environment where the crops are cultivated. The IoT was utilized for gathering data from environmental attributes: air temperature and humidity, carbon dioxide concentration, soil moisture and temperature, and light intensity. The data from IoT are then analyzed through different AI models for adjusting the environment for crop growth [76]. Another hydroponic setup for lettuce production utilizes an automatic control of pH and nutrient solution concentration. The system used sensors for gathering data, microcontrollers for data processing and actuators for responding based on the results of AI algorithm to effectively adjust the nutrient parameters autonomously [77]. These studies applied AI for AMS in a CEA.

A machine vision system (MVS) was implemented for automatic classification of different leaves if they have defects or diseases. For this MVS, Haralick algorithm was used for extracting texture features. The features build the dataset for training different ML algorithms: artificial neural network, Naïve Bayes, random forest, and support vector machine; to meet the objective in determining if leaves are in good condition or not [57].

It is established that controlled environment agriculture is one of the most common applications of innovative urban agriculture (IUA), which uses the discipline of biosystems engineering for ensuring efficient crop production and energy consumption while considering environmental sustainability. Among the various CEAs, aquaponics is the most relevant research focus on the recent years. However, there are research gaps involving AP performances that need to be filled as this system has numerous challenges as it involves cultivars of different species sharing the same medium for nutrient consumption. One of the pressing issues in an AP system is the difficulty to maintain recirculating water that provides the right number of substances for both fish culture and the vegetable crop. Therefore, controlling these kinds of parameter significantly contributes for the effectiveness of such systems. Agricultural automation enables remote controlling and monitoring, which eliminates the need to have complex procedures for maintaining a suitable environment for growth. Moreover, agricultural automation for a smart aquaponics system can perform better when considering an adaptive management system (AMS), enabling the automation to adjust the environmental factors affecting cultivation based on the real-time condition and status of the cultivars. In this way, optimum results can be achieved in terms of overall success determinant. An adaptive AP system is done by integrating artificial intelligence or computational intelligence as it is capable to respond accurately and immediately as it pre-learned the situations or circumstances in the ecology.

3. Processes involve in formulating computational intelligence-based adaptive management system for urban agriculture (CIAMSUA)

Figure 1 shows the conceptual framework of the proposed CIAMSUA aquaponics platform. The framework consists of five phases on which each phase composed of detailed methods for integrating from the software to the hardware prototype. Phase 1 involves the implementation of sensor nodes to the four environmental systems. The first phase includes calibrating the sensors, programming them for data acquisition, and design and development of sensor node implementation. The second phase is the development of control and automation through programming of the actuators based on logical response from data and the construction of systems for the control and automation processes. Data transmission and acquisition comprise phase three, it integrates and embeds the wireless communication programs in the microcontroller nodes. As these data are transmitted to the cloud, it will be effectively acquired from the cloud for data processing. Data from cloud will not be limited from the sensors, which would also include data from the machine vision systems. Modeling the performance of controls based on production is done at the fourth phase. This is going to be implemented though training machine learning algorithms and adapting them in the system.

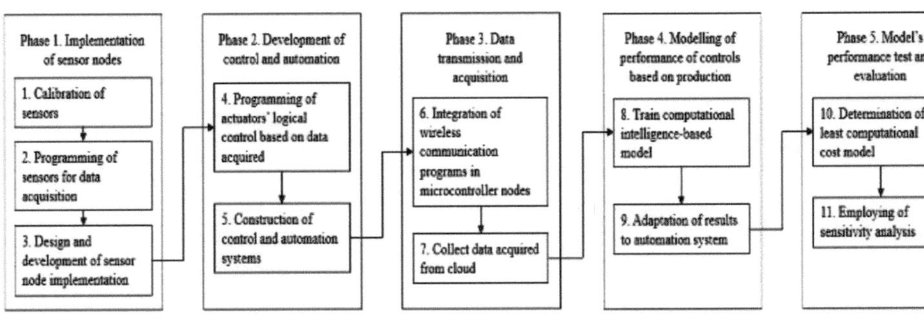

Figure 1.
Developmental framework of computational intelligence-based adaptive management system for urban agriculture (CIAMSUA).

3.1 Sensor programming for data acquisition

Data read and acquired from sensors may vary even if they were produced with the same manufacturers. There could be slight differences in the values yielded for each sensor detecting the same parameter and environment, which leads to inaccuracy and inconsistency. This is very critical and significant to consider as computational intelligence heavily rely on data for training to effectively perform. To accurately gather data, sensors are calibrated in the same external factors to adjust their readings. After calibration, sensors are programmed to acquire or read data where they are deployed. In this research, Arduino integrated development environment (IDE) will be used for embedding the software codes to the hardware through a microcontroller. Specifically, the code will be written in C ++ as this is the language used by the Arduino IDE.

The code will be composed of initiating libraries to efficiently apply existing functions so the program can simply call the specific operation to logically provide the responses. Pin configurations will be setup for assigning sensors to which pin in the microcontroller will be connected. Variables will then be initialized depending on what type (i.e., integer, float) of data they are. Void setup will then be programmed to activate variables as pins and determine which pins will be used as an input or output mode. This will also include initiating serials and sensor reading operations. The void loop will be written with sensor reading programs for the different environmental factor systems. The code will then be embedded to the microcontroller to a universal serial bus (USB). Specifically, shown in **Figure 2**, ESP32 will be used as a microcontroller as it has a Wi-Fi module integrated with the chip at an inexpensive price. Data to be acquired will be from four different systems of CIAMSUA. The following are the list of environmental factors to be monitored:

1. For nutrient mixture, conditioning, and drain tank automation,

    a. Power of Hydrogen (pH) Level

    b. Electrical Conductivity (EC)

    c. Dissolved Oxygen (DO) Level

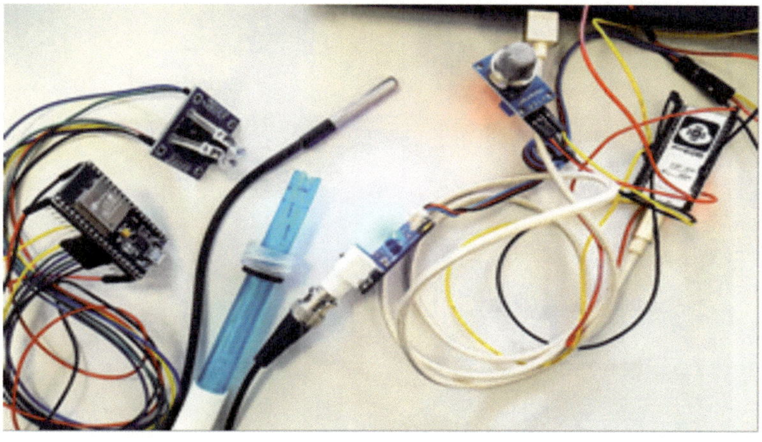

Figure 2.
Programming sensors for data acquisition with ESP32 board.

d. Ammonia (NH3) Level

e. Water Temperature

2. For irrigation control,

a. Water level in a tank

3. For artificial lighting systems,

a. Real-time basis

4. For temperature control,

a. Environment temperature

b. Environment humidity

The CIAMSUA will be focused and implemented in the hydroponics chamber platform. The irrigation control concentrates on managing the water from fishpond to be properly distributed in the chambers for the hydroponics unit to effectively produce lettuce crops.

3.2 Sensor node implementation

The schematic diagram shown in **Figure 3** is one of the sensor nodes for the nutrient mixture automation, the conditioning tank, and the irrigation control.

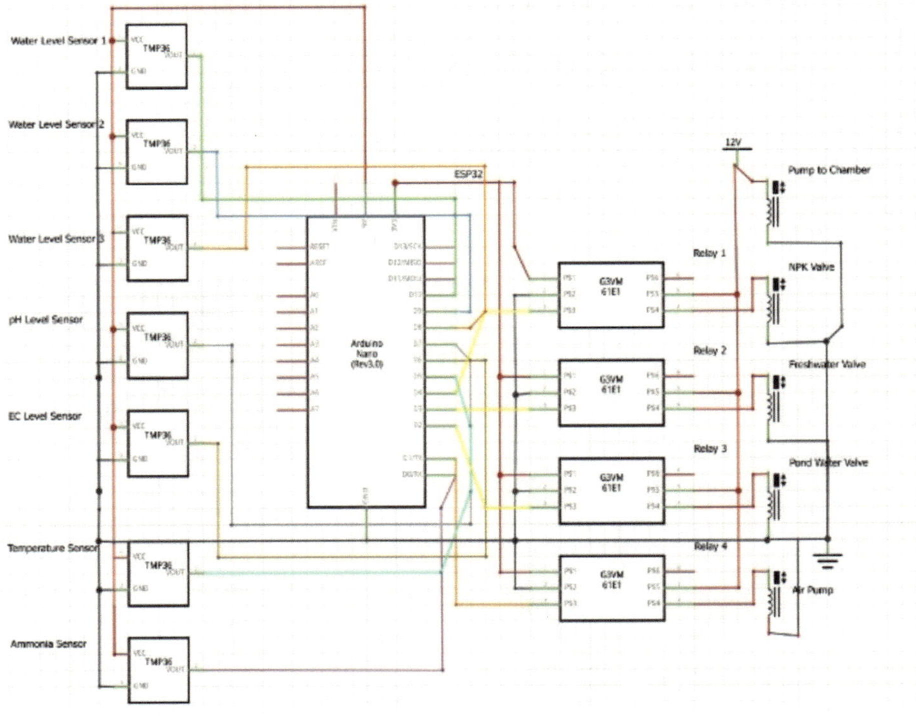

Figure 3.
Nutrient mixture, conditioning tank, and irrigation control node schematic acquisition.

There are three water-level sensors placed at the input side. Four more sensors are also assigned as input, and these are pH level sensor, EC level sensor, water temperature sensor, and ammonia sensor. Logical controls are embedded on the ESP32. The outputs are connected to an electronic mechanical relay, which response to trigger the microcontroller to switch on and off the solenoid valves, the water pump, and the air pump for appropriate water flowing and mixing.

Figure 4 shows the schematic diagram of the node to be placed in the drain tank. As noticed, there are no actuators in the node as the purpose of this is only to acquire data to determine the difference between the limnological parameters in the water before and after the lettuce crops consumed the water being irrigated in the racks of the chambers. The sensors included are the same as the sensors shown in **Figure 3**; that way, a model can be derived to determine how much nutrients of the crops do intake in different life stages.

Figure 5 shows the node that both control the NPK solution tank and the freshwater tank. The control is done for allowing the liquid to pass through the hose leading to the solenoid valves connected in the nutrient mixture and the conditioning tank. The sensors are not connected to the control and will only be used for monitoring the tank levels so that a manual refill can be made if the tanks are empty and manual drainage if the tanks are full.

Sensor node for temperature control is shown in **Figure 6**. It has three temperature sensors for acquiring real-time temperature in different areas of the hydroponics chamber. The output side is a relay that is electronically triggered to switch on or off the exhaust fan. The relay also functions to isolate the low DC voltage of the input and microcontroller side from the high AC voltage of 220 V that is required to operate the exhaust fan.

The artificial lighting system schematic diagram is shown in **Figure 7**. The ESP32 is not connected to any input devices as this automation depends on time duration. A delay function would be embedded to switch on or off the relays connected to lights at a specific amount of time.

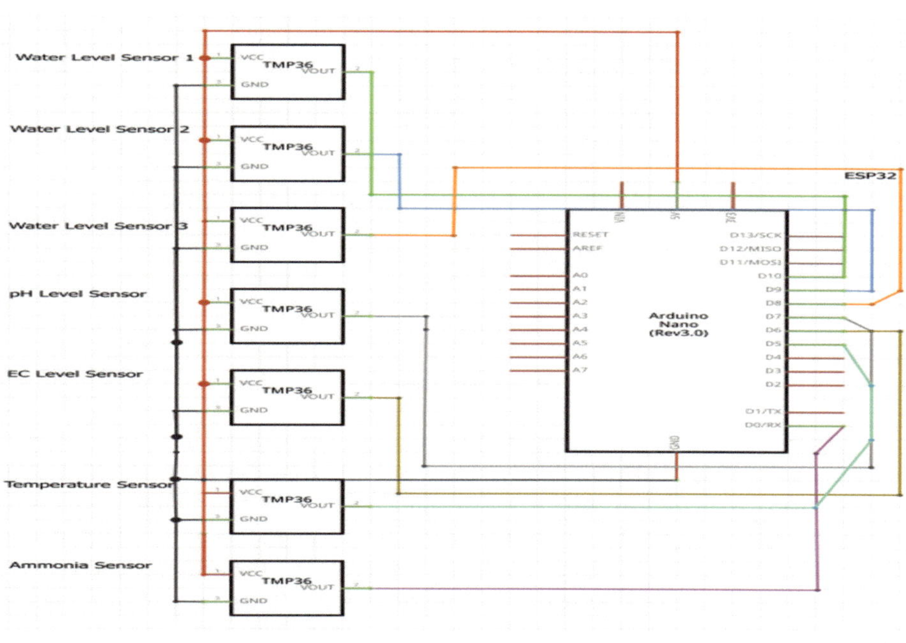

Figure 4.
Drain tank control node schematic diagram.

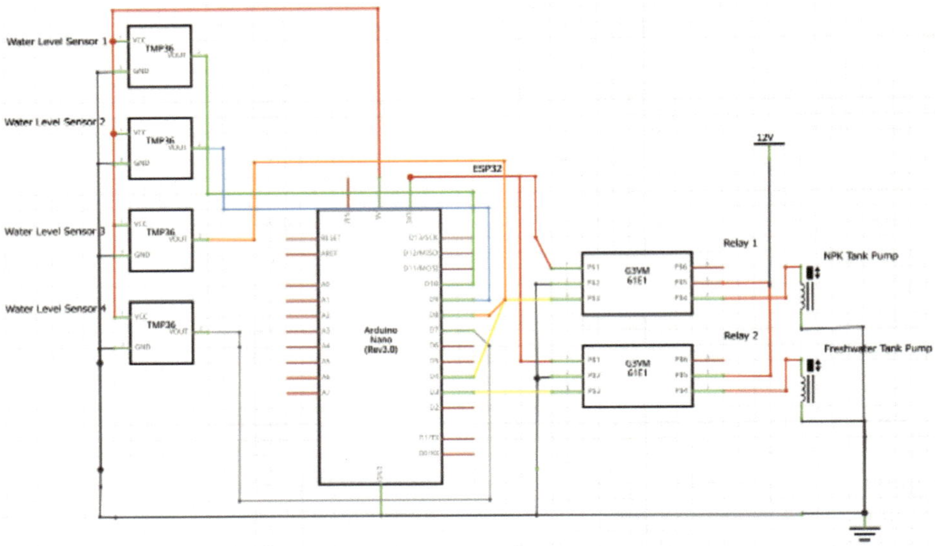

Figure 5.
NPK and freshwater tank control.

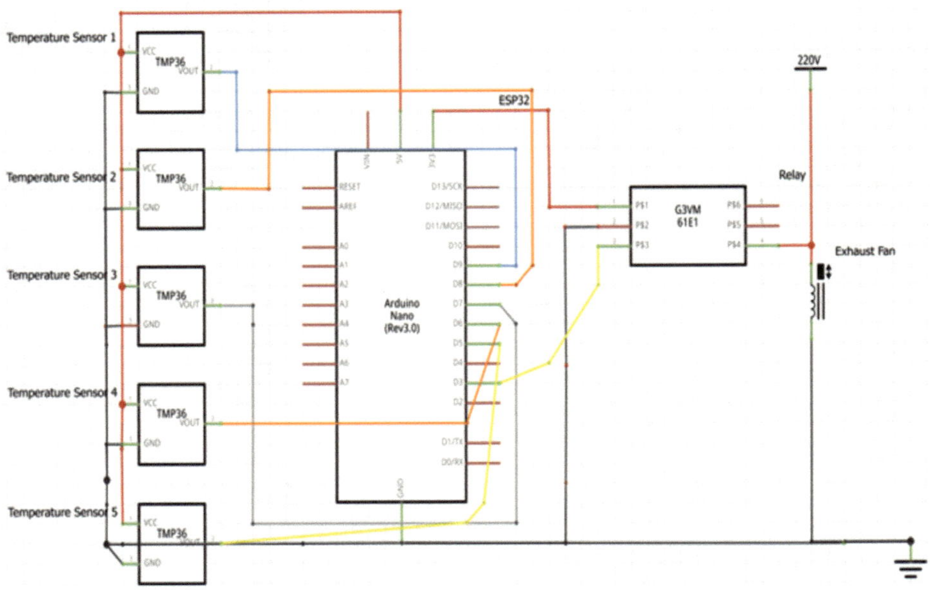

Figure 6.
Temperature and humidity maintenance node schematic diagram.

Actuators will be logically programmed in accordance with the data acquired from the sensors. The truth table only presents the system for the nutrient mixture automation and irrigation control as shown in **Table 1**. The truth table developed to visualize all the eight possible combinations of water-level sensors status and the corresponding responses of the five actuators involved in the nutrient mixture automation and irrigation control. Out of eight combinations, only four are realistically possible as the water-level sensors are placed in different levels in the water tank, resulting to only considering combinations that correspond to water filling from the bottom of the tank up to the top consecutively.

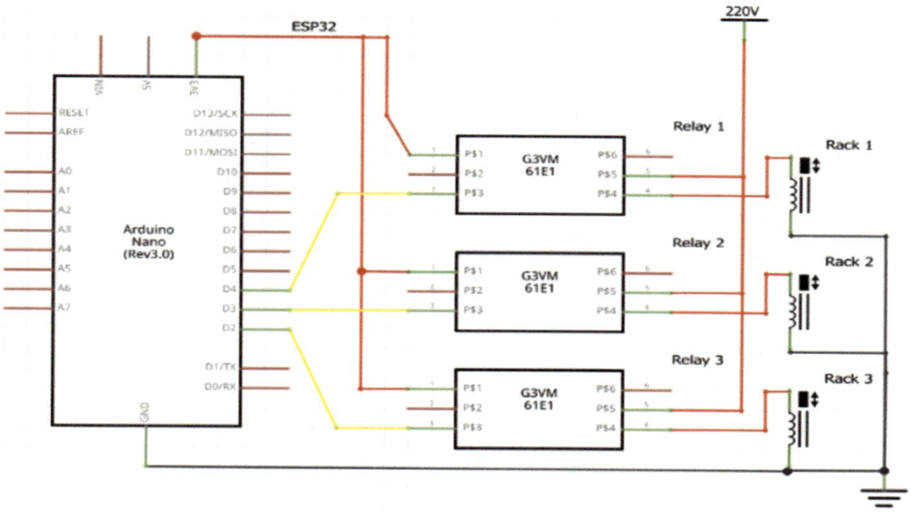

Figure 7.
Artificial lighting system node schematic acquisition.

The first combination, 0–0-0, means that the water in the tank is at the bottom level, not reaching any of the three sensors. With this, the pond water valve and the pump supplying water to the chamber from the tank should be on. At 0–0-1, the water level is being detected by the sensor placed at the lowest level. The 0-1-1 combination shows that the two sensors placed at the bottom and at the middle of the tank are turned on, representing that the tank is half full; thus, valve for pond water should be turned off to avoid overfilling. The top half of the tank should only be filled with either freshwater or NPK solution to control the pH and EC level of the tank mixture. The combination 1–1-1 means that the water sensor at the top-most level of the tank has been reached, remarking it as full; therefore, the source water from the pond should be stopped. The combination 0–1-0 is not possible to take place as the water cannot only be at the middle part of the tank and not at the bottom. The same is true for the three remaining truth table combinations.

Readings from pH, EC, temperature, and ammonia sensors should always be activated, representing a logic 1 to continuously acquire data at any water level. The pump-distributing water to the chambers should always be turned on as well at any level, to consistently supply water to the lettuce racks. Air pump on the other hand follows an OR logic, on which it will turn on only if one of the valves is on.

3.3 The CIAMSUA fuzzy logic controller

Figure 8 shows the fuzzy logic controller (FLC) that is use in this study. Data that were transmitted to a cloud-based dashboard will be automatically downloaded in MATLAB. From there, they would be used as new dataset for the pretrained network that is connected to the FLC. The output of the computational intelligence-based models, which are specifically the fresh weight and the phytopigments of the lettuce crops, will be used as input to a fuzzy inference system for determining how long will the NPK valve, freshwater valve, and air pump should turned on or off. The truth table and the FLC function together as a nested condition of the actions for irrigation controls to determine the control of the air pump, NPK, freshwater, and pond water valves.

The membership functions of the input and output for the fuzzy logic control of the freshwater and NPK solenoid valves are shown in **Figure 9**. The fresh weight of

	Tank status			Remarks				Actuators' Response				
H_2O level top	H_2O level middle	H_2O level bottom	Remarks	pH	EC	Temp	NH_3	NPK valve	Fresh H_2O valve	Pond H_2O valve	Pump to chamber	Air pump
0	0	0	No to very low level water	1	1	1	1	0 or 1	0 or 1	1	1	0 only if NPK and fresh H_2O valves are 0
0	0	1	Water level at lowest sensor (3)	1	1	1	1	0 or 1	0 or 1	1	1	0 only if NPK and fresh H_2O valves are 0
0	1	0	Cannot happen	—	—		—	—	—	—	—	—
0	1	1	Tank halfway full	1	1	1	1	0 or 1	0 or 1	0	1	0 only if NPK and fresh H_2O valves are 0
1	0	0	Cannot happen	—	—	—	—	—	—	—	—	—
1	0	1	Cannot happen	—	—	—	—	—	—	—	—	—
1	1	0	Cannot happen	—	—	—	—	—	—	—	—	
1	1	1	Tank full	1	1	1	1	0	0	0	1	0 only if NPK and fresh H_2O valves are 0

Table 1.
Irrigation control truth table.

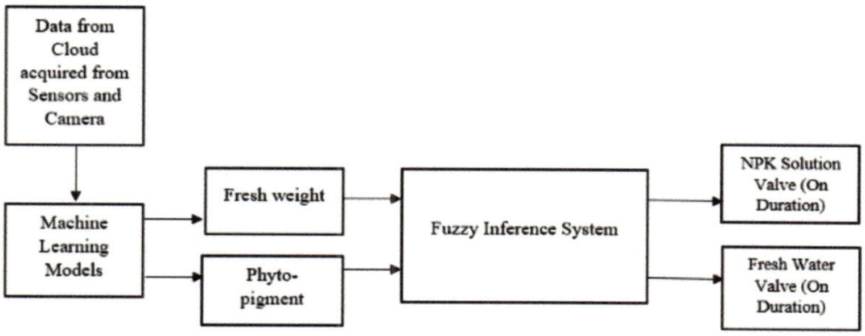

Figure 8.
Nutrient mixture tank fuzzy logic controller.

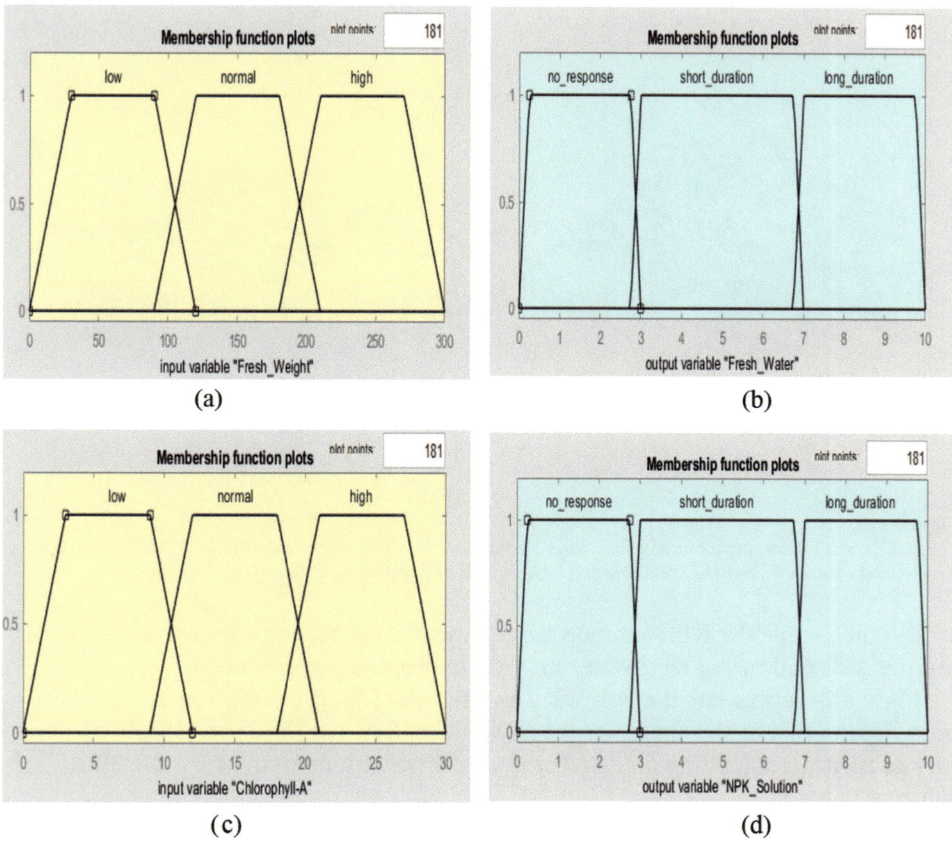

Figure 9.
(a) Fresh weight, (b) freshwater, (c) chlorophyll-A, and (d) solenoid valves membership functions.

the lettuce is used as input in **Figure 9a** representing the 0 to 300 mg range of change in weight per day. **Figure 9c** represents the membership function for the 0 to 30 mg/L range of changes of chlorophyll-A per day. The output crisps represent the duration for how long the valves: freshwater **Figure 9b** and NPK solution **Figure 9d** are on, which ranges from 0 to 10 seconds.

The rule viewer for the fuzzy inference system (FIS) is shown in **Figure 10**. Based on the given sample values of the input parameters, having 269 mg of change in fresh weight in a day and 23.2 mg/L change of chlorophyll-A in a day, the freshwater valve will turn on for 8.37 seconds. Another system was implemented to round it off to

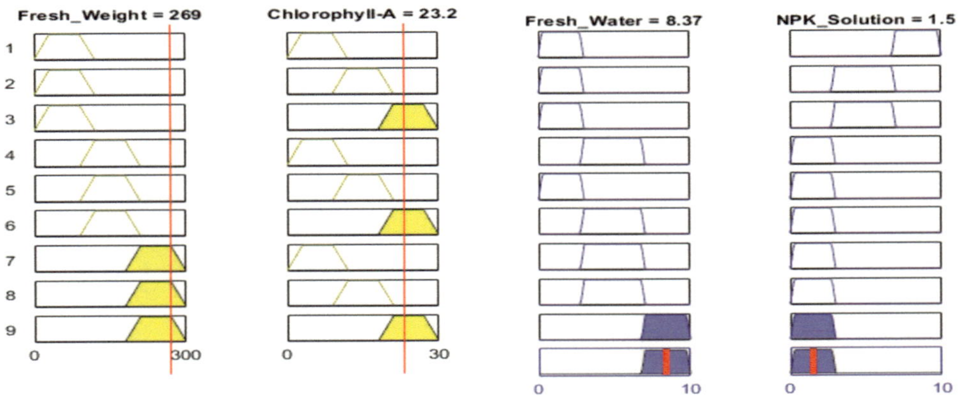

Figure 10.
Fuzzy inference system rule viewer with nine rules.

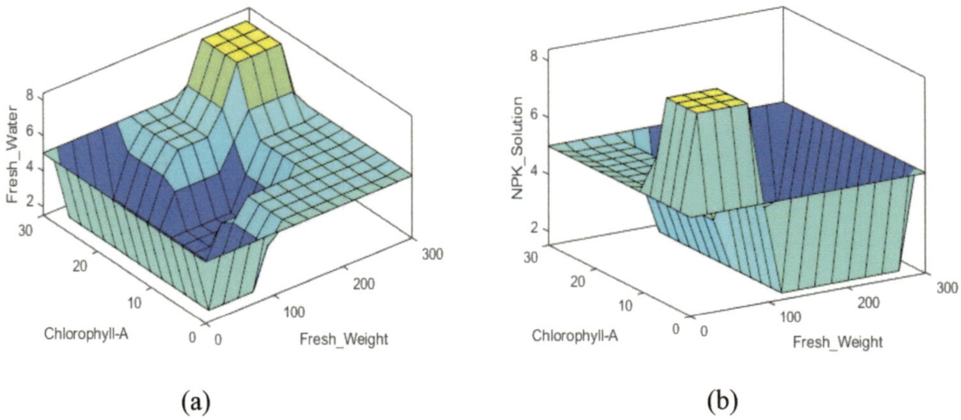

Figure 11.
Fuzzy inference system surface viewer for lettuce phenotypes: (a) fresh water, chlorophyll a and fresh weight interrelated dynamics; and (b) NPK solution, chlorophyll and fresh weight interrelated dynamics.

10 seconds, while the NPK solution valve is turned off for the day's cycle. Thus, this sample adds 660 mL of freshwater into the hydroponics growth bed.

Figure 11 represents the Surface Viewer of the FIS. It can be interpreted from both figures that as the fresh weight and chlorophyll-A values increase, it is necessary to turn the solenoid valve on for the freshwater longer and turn the NPK solution valve off.

The inference systems using freshwater and the other two phytopigments: chlorophyll-B and vitamin C follow the exact same membership function, ranges, and sets of rules used. Thus, only one set of FLC is represented.

The artificial lighting systems will be controlled depending on photoperiod, which is reliant on a time duration input. Various durations will be assigned for each layer of lettuce rack for experimentation. By default, the test cases will be set to 9, 12, and 16 hours, respectively. This will then change depending on the results of the performance evaluation or sensitivity analysis from the output of the models developed so that permanent photoperiod control could be integrated in the system.

The temperature is programmed through a fuzzy logic controller to maintain the temperature adequate for the plant's excellent growth. **Figure 12** shows the block diagram for the adaptive control of the temperature and humidity maintenance based on the fresh weight and phytopigments of lettuce crops. The system follows the similar approach with the nutrient mixture control. However, there is a

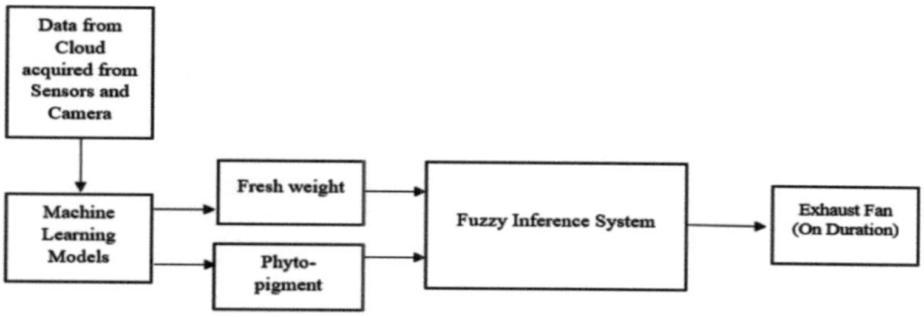

Figure 12.
Temperature fuzzy logic controller.

significant difference on how it was implemented. For the temperature and humidity control, the FLC was embedded to the microcontroller through the Arduino IDE, unlike with the nutrient mixture control which uses the Simulink.

Listed below are the actuators that will be implemented for each of the four systems:

1. For nutrient mixture tank automation,
    a. Freshwater solenoid valve.
    b. NPK solenoid valve.
    c. Freshwater solenoid valve.
    d. Air pump
2. For irrigation control,
    a. Water pumps
3. For artificial lighting systems,
    a. Photoperiod control
4. For temperature control,
    a. Exhaust fan.

3.4 Data acquisition for CIAMSUA

The aquaponics system was constructed following the layout in **Figure 13**. The control nodes were placed accordingly to where it is needed the nearest to avoid long-wiring connections. **Figure 13a** shows the front view of the hydroponic chamber on which the drain node, artificial lighting system, and temperature and humidity controls are visible. **Figure 13b** shows the back of the hydroponic chamber on which the fishpond for the aquaculture unit is located. Two nodes were placed, which are the nutrient mixture and the NPK solution tank nodes. **Figure 13c** and **d** shows additional isometric views that further provide understanding on where the nodes were implemented and placed in the chamber. The 3D model was only limited to represent the adaptive management system. **Figures 14–18** show how the hardware nodes for the automation are constructed in terms of wiring

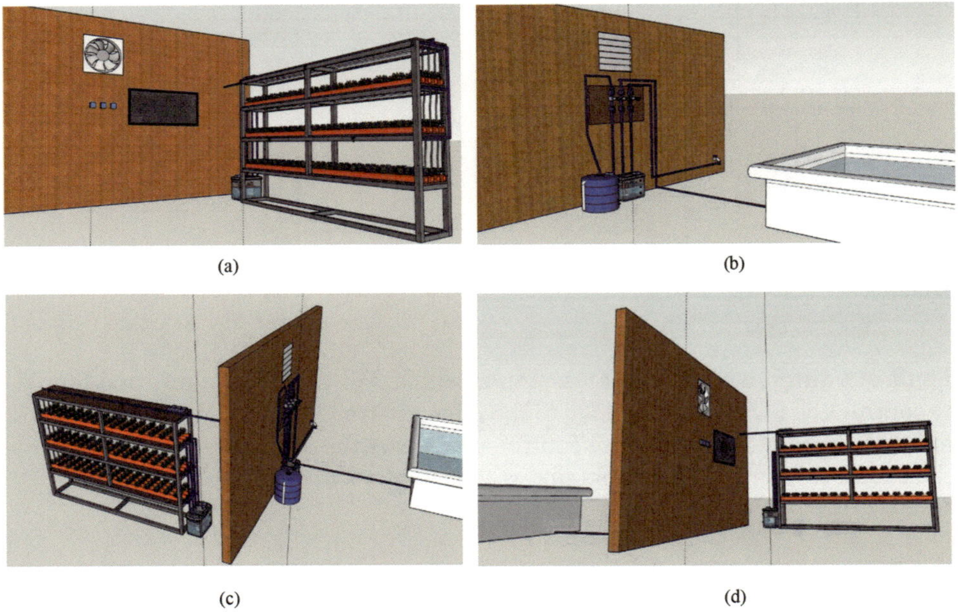

Figure 13.
Hardware implementation of the aquaponics setup: (a) front view of the aquaponic chamber with emphasis on the installed exhaust fan and grow bed, (b) mixing tank connected to an artificial fish pond, (c) drain tank connection from the vertical grow bed, and (d) connection from artificial fish pond to the grow bed.

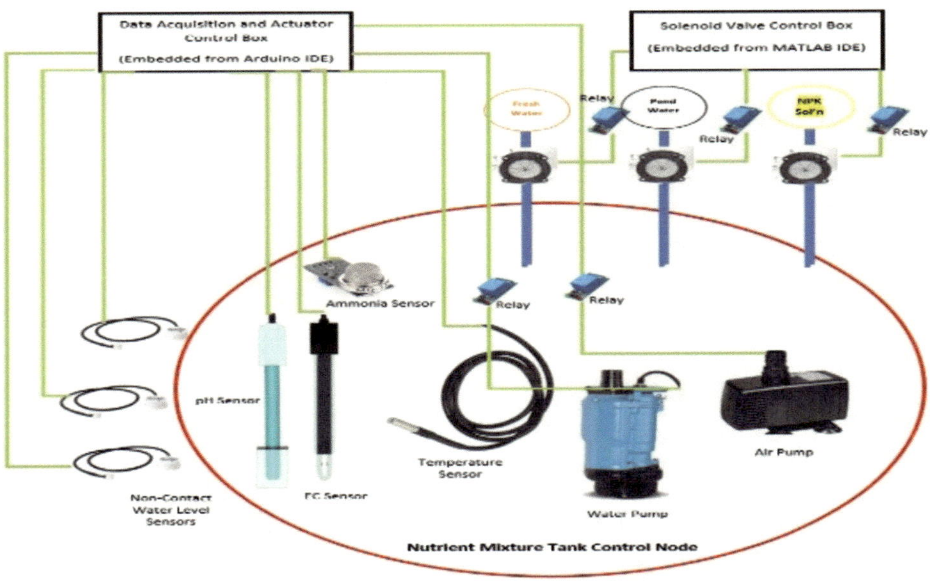

Figure 14.
Nutrient mixture tank control node.

diagram as reflected in the schematics. The details of its programs are already discussed in the previous sections.

Aside from controlling the actuators, monitoring of the sensor and actuator status was implemented to determine whether the acquired data match the actuators responses. The status of each device, whether they are active or not, was transmitted alongside with the data from the sensors to the cloud to a wireless network.

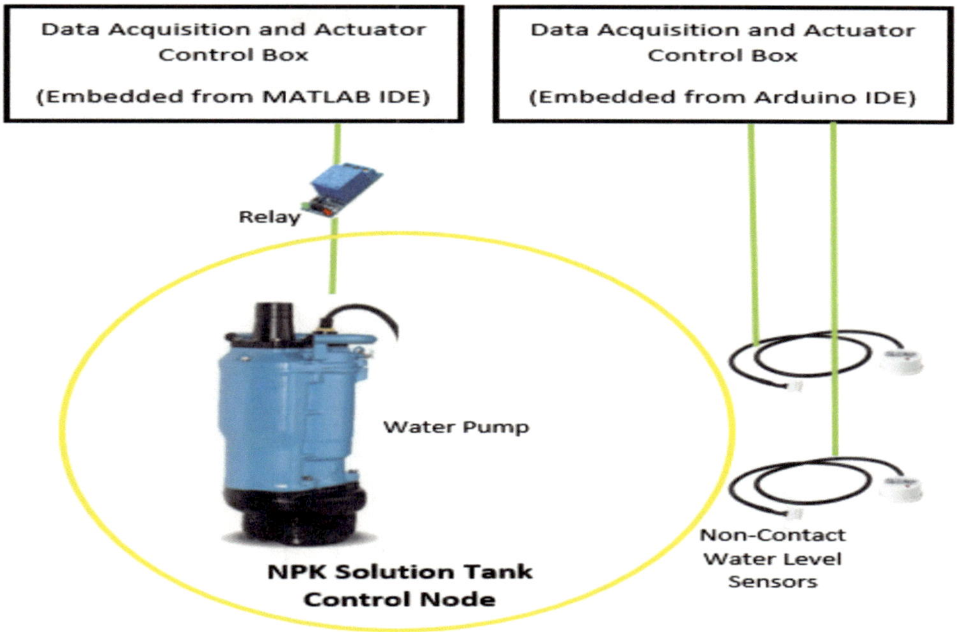

Figure 15.
Freshwater and nutrient solution tank control node.

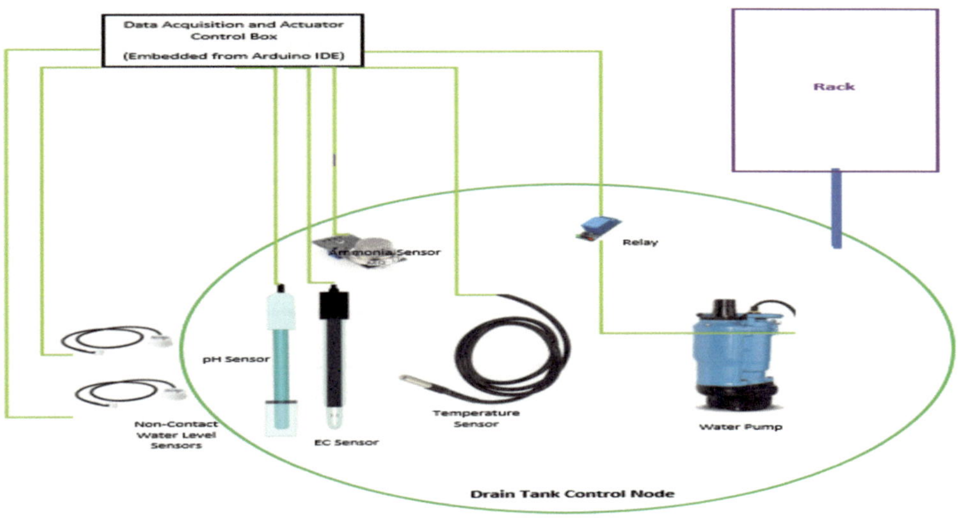

Figure 16.
Drain tank control node.

The data acquired from the sensors were transmitted *via* wireless communication network. This is done by embedding programs that activate the Wi-Fi module of the ESP32 and selectively send the data to a common router. The common router is responsible for collectively sending data to the cloud. Data will be collectively displayed on a cloud dashboard for remote monitoring. The data printed from the dashboard will be exported to csv files to present a tabular form of dataset. **Figure 19** shows a sample monitoring dashboard for displaying the data acquired.

Spectrophotometry was done to obtain the response variables that are the chlorophyll-A, chlorophyll-B, and vitamin C for a given input image. This is the technique used to carry out the discipline of spectroscopy. It is a method for

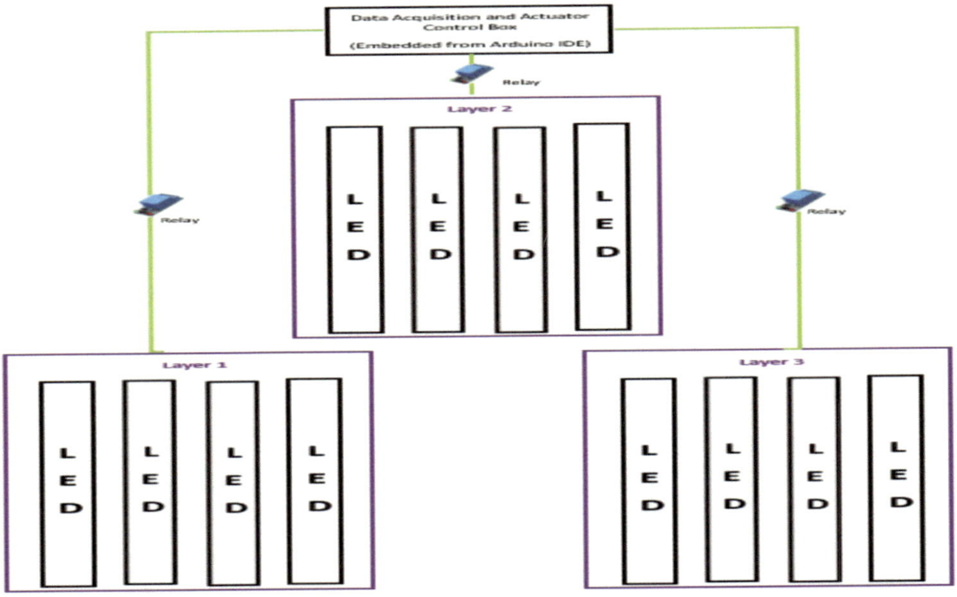

Figure 17.
Artificial lighting system control node.

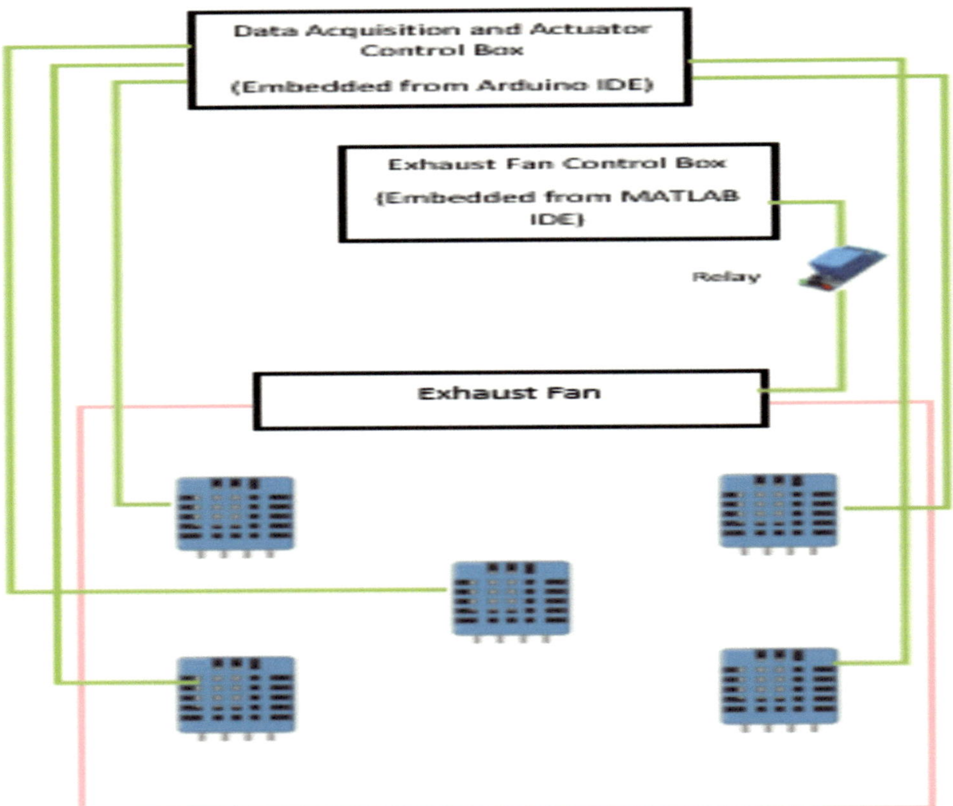

Figure 18.
Temperature control node.

quantitatively measuring the light spectra reflection as dispersed in the concept of spectroscopy and its interaction to the properties of materials' transmission relative to the wavelength. It measures the light's relative intensity at a specific wavelength,

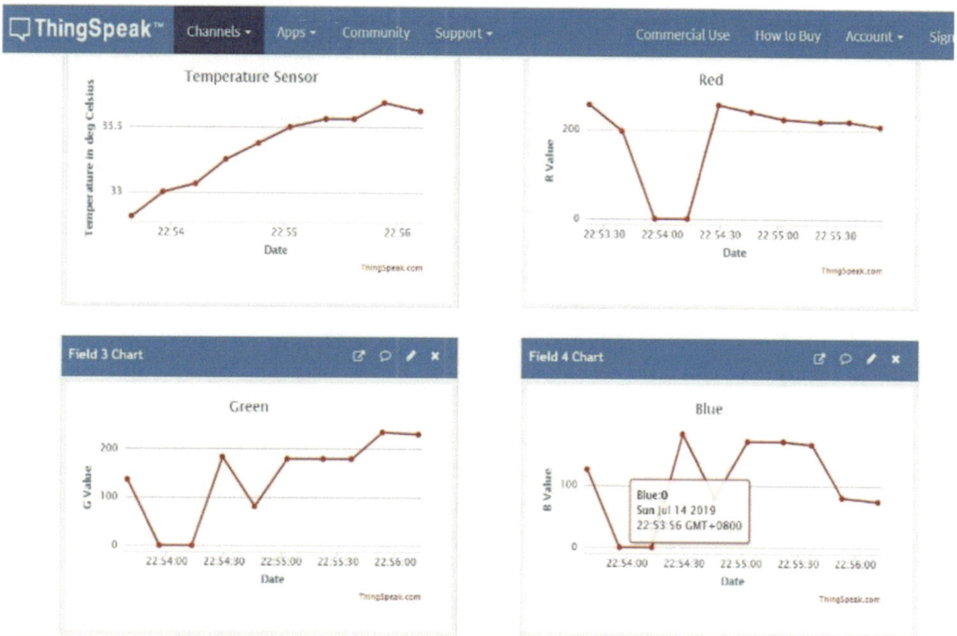

Figure 19.
Sample cloud dashboard for data acquisition.

(a) (b)

Figure 20.
Sample test tubes during pigment extraction from (a) leaf tissue samples mixed with ethanol and (b) collection of test tubes containing different concentrations of dissolved leaf pigments.

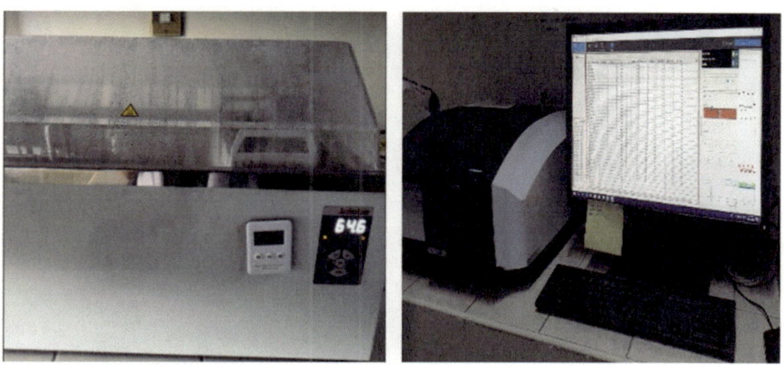

(a) (b)

Figure 21.
(a) Hot bathing and (b) spectrophotometry for pigment assay.

which supports spectroscopy on determining the relationship of the spectrum absorption properties of substances. **Figure 20a** and **b** show the test tubes that contain a leaf sample for every lettuce per day. These were then heated at 65°C as shown in **Figure 21a** before undergoing spectrophotometry in **Figure 21b**.

Image processing was done to resize the images to the compatible input sizes of a given transfer learning network. These transfer learning networks were used in predicting the fresh weight, chlorophyll-A, chlorophyll-B, and vitamin C given a lettuce image as input. InceptionV3 were used for predicting the fresh weight, chlorophyll-B, and vitamin C, while MobileNetv2 was used in predicting chlorophyll-A. For Inceptionv3, the required image size is 299x299x3 pixels, while for the MobileNetv2 it is at 224x224x3 pixels. Hence, the captured images were resized to those values. These networks were preselected based on multiple trial and error on determining which networks would yield the highest accuracy in performing the regression. **Figure 22a** display the original image captured, while **Figure 22b** is a sample of an image resized to 299x299x3 to be used as input in transfer learning. The output on the other hand is the corresponding data gathered from spectrophotometry. The data created were split into 70% training datasets and 30% testing datasets. The training dataset was used to model four regression networks.

3.5 CIAMSUA model performance

The modeled algorithms were evaluated to determine its performance through the "unseen" data or the testing data to validate if the models do not overfit to the training data alone. **Figure 23** shows the transfer learning network. The first dataset is bigger than the dataset used in the study as it would be used to pre-train a more generalized model. The target domain in the study used segmented lettuce images. The source model transfers knowledge to a target model to perform prediction of phyto-morphological features. The models were developed while considering hyperparameter optimization to further increase the algorithm's performance. Training options were predetermined and set before training the model. The parameters and their values are as follows:

1. Optimizing algorithm—Stochastic Gradient Descent with Momentum

2. Mini Batch Size—32

(a) (b)

Figure 22.
Sample input image (a) at t_1 = 0 hr and (b) at t_2 = t_1 + 24 hrs.

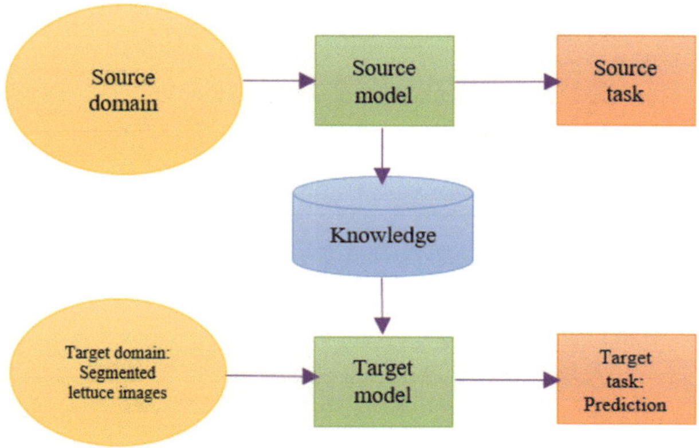

Figure 23.
Transfer learning network.

3. Initial Learning Rate—0.000001

4. Maximum Epochs—10

After training the models, a complete cultivation of lettuce was conducted on which images were captured each day for the six-week cultivation. The capture images were used as input to predict the fresh weight and the three phytopigments.

Table 2 shows the predicted values of the fresh weight and the three phytopigments daily. The average of the dataset was obtained which served as the ground truth for the standard values for the fresh weight, chlorophyll-A, chlorophyll-B, and vitamin C as summarized in **Table 3**. The values in the table were also fed back to the fuzzy logic controller to be used as parameters for the input of the inference system. The change of values per day was obtained to be set as ranges for the membership functions.

Evaluating the system's performance will be reliant on machine vision algorithms and the data acquired from sensors to determine the quality of the crops produce. From the results of quality assessment, it will determine the response of the automation and control system to produce the environmental factors need by the crops to improve their quality. This therefore adjusts the nutrients being supplied to the crops, the pressure of water flow feeding the crops, the amount of light intensity and photoperiod, and the temperature of the surroundings resulting to an adaptive and managed system. The system adaptation will be done by integrating MATLAB, which is capable of machine vision and machine learning algorithm to be embedded in the microcontroller.

To further improve the system models developed, least computation cost was implemented. This finds out which systems have the least mean-squared error (MSE) and least learning time for training while maintaining a high performance. **Figures 24–26** are examples of the adjustment that were conducted for the number of neurons for determining which model would yield the least MSE, least-learning time, and high accuracy.

Sensitivity analysis is a way to determine how the dependent variables, which are the fresh weight and phytopigments, are being affected by the independent variables that are the amount of NPK solution, freshwater, environmental temperature and humidity, and the photoperiod control. The experiment for this phase would be to measure the amount of NPK solution and freshwater for the short and long duration as produced by the fuzzy logic controller (FLC). The process also

Day	Image no	Fresh weight (g)	Chl-a (mg/L)	Chl-b (mg/L)	Vit C (mg/L)
1	1	0	87.221	135.623	290.369
2	2	0.45	97.143	157.623	336.39
3	3	0.52	100.449	174.236	369.125
4	4	0.78	139.874	189.623	377.36
5	5	0.84	147.869	209.143	382.898
6	6	0.856	157.41	220.32	387.27
7	7	0.88	167.84	226.52	395.369
8	8	0.906	167.902	234.63	397.739
9	9	0.914	169	236.041	400.321
10	10	0.925	169.227	241.356	401.376
11	11	0.937	170.914	248.73	402.36
12	12	0.97	171.265	253.559	403.013
13	13	1.019	173.415	265.325	405.694
14	14	0.986	175.347	271.143	410.364
15	15	1.03	177.846	285.694	418.395
16	16	1.167	191.623	297.137	426.375
17	17	1.354	215.412	309.67	439.657
18	18	1.497	229.654	327.65	444.36
19	19	1.524	245.562	350.571	456.949
20	20	1.74	257.027	376.197	468.791
21	21	1.929	278.974	387.462	479.765
22	22	1.974	282.475	394.273	481.367
23	23	2.004	294.613	401.657	487.673
24	24	2.194	301.214	415.193	492.317
25	25	2.201	303.415	436.874	495.769
26	26	2.313	309.919	468.198	498.185
27	27	2.461	314.95	491.257	507.769
28	28	2.61	321.54	505.056	519.469
29	29	2.899	347.618	516.243	523.076
30	30	3.164	374.512	547.35	531.176
31	31	3.334	391.147	587.367	542.176
32	32	3.447	407.874	598.38	554.007
33	33	3.656	417.843	617.978	561.073
34	34	3.779	431.512	628.367	570.312
35	35	3.962	467	654.328	587.675
36	36	4.219	498.184	690.143	668.447
37	37	4.675	502.634	739.165	708.98
38	38	4.914	520.21	784.361	781.142
39	39	5.389	529.347	800.35	815.6
40	40	5.759	564.955	816.646	831.194

Day	Image no	Fresh weight (g)	Chl-a (mg/L)	Chl-b (mg/L)	Vit C (mg/L)
41	41	6.119	575.695	837.361	854.989
42	42	6.221	592.314	865.32	872.994

Table 2.
Fresh weight and phytopigments.

STANDARD				
Week	Freshweight (g)	Chl-a (mg/L)	Chl-b (mg/L)	Vit C (mg/L)
1	0.618	128.258	187.584	362.683
2	0.951	171.01	250.112	402.981
3	1.463	228.014	333.483	447.756
4	2.251	304.018	444.644	497.507
5	3.463	405.358	592.859	552.785
6	5.328	540.477	790.478	790.478

Table 3.
Phytopigments standard values of lettuce per week.

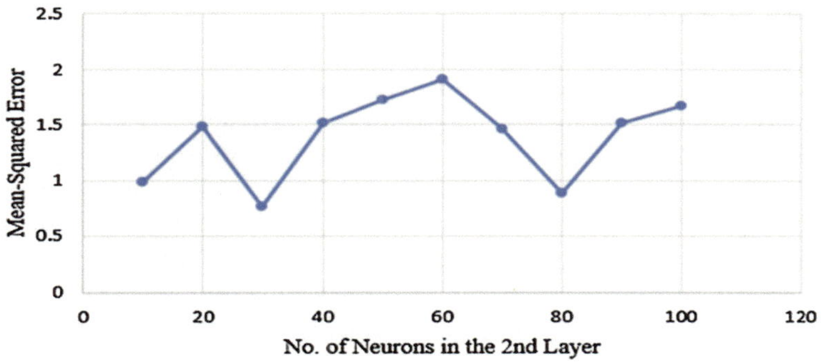

Figure 24.
Number of neurons versus mean-squared error.

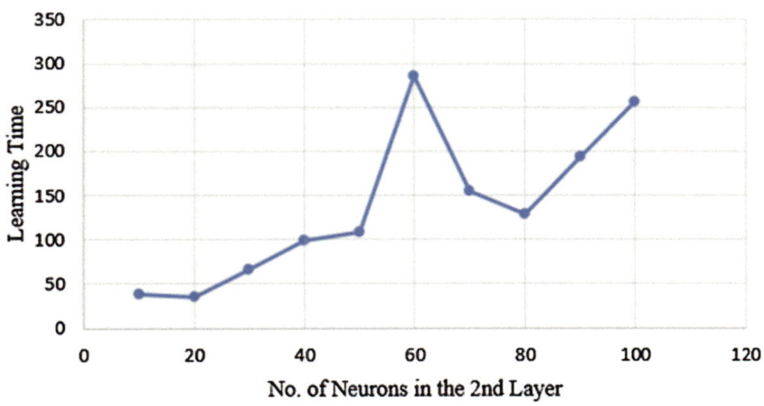

Figure 25.
Number of neurons versus learning time.

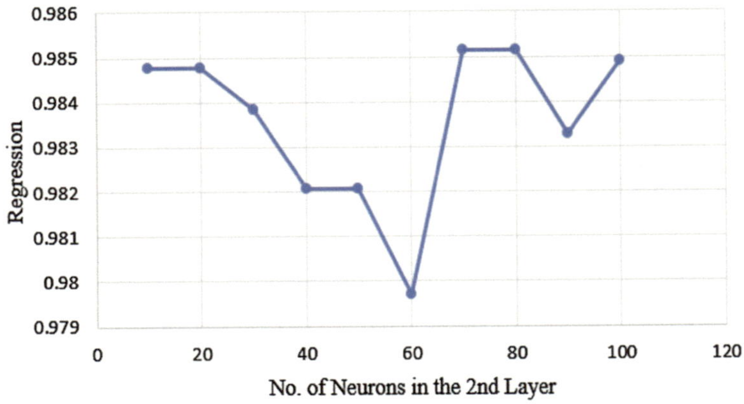

Figure 26.
Number of neurons versus regression.

includes the following: a) determine the temperature, humidity, and the corresponding photoperiod of artificial lights, b) measure crop parameters, c) record the FLC conditions and crop parameters for the whole crop cycle, and d) determine whether crop features improve from the given FLC condition.

4. Results and discussion

This chapter details the results of the experimentation conducted as followed from the methodology and the discussions for supporting the validity of achieving the objectives. For the nutrient mixture, conditioning tank and irrigation control, sensitivity analysis was done to determine the response of the control system based on the data acquired from lettuce images. The artificial lighting system was experimented with the aid of the test case from **Table 4**. The temperature and humidity maintenance node was tested by determining the response of the fuzzy logic controller given with simulated environment temperature and humidity. Automating the system adaptively follows standard target values to provide the necessary amount of nutrients that are needed to sustain the standard phytopigment values for optimal lettuce crop growth. The lettuce crops, fresh weight, chlorophyll-A, chlorophyll-B, and vitamin C should be at least 5.328 g, 540.48 mg/L, 790.48 mg/L, and 790.48 mg/L, respectively, during the harvesting stage at week six.

	Rack with 9 hours of on state in a photoperiod	Rack with 12 hours of on state in a photoperiod	Rack with 16 hours of on state in a photoperiod
Fresh weight on Day 42	4.923 g	5.328 g	5.013 g
Chlorophyll-A on Day 42	481.914 mg/L	540.477 mg/L	545.896 mg/L
Chlorophyll-B on Day 42	619.847 mg/L	790.48 mg/L	708.367 mg/L
Vitamin C on Day 42	611.598 mg/L	790.48 mg/L	756.284 mg/L

Table 4.
Phytopigments values during harvest stage on different cultivation photoperiods.

DOI: http://dx.doi.org/10.5772/intechopen.102775

	Freshwater										
Day	**0**	**1140**	**2280**	**2760**	**5220**	**5700**	**7320**	**7800**	**8940**	**9420**	**10,080**
1	0	0.052946	0.105892	0.128185	0.203424	0.225717	0.317675	0.362261	0.415207	0.4375	0.468153
2	0.024974452	0.07792	0.130866	0.153159	0.228398	0.250691	0.34265	0.387236	0.440182	0.462475	0.493128
3	0.049948903	0.102895	0.155841	0.178134	0.253373	0.275666	0.367624	0.41221	0.465156	0.487449	0.518102
4	0.074923355	0.127869	0.180815	0.203108	0.278347	0.30064	0.392599	0.437185	0.490131	0.512424	0.543077
5	0.099897807	0.152844	0.20579	0.228083	0.303322	0.325615	0.417573	0.462159	0.515105	0.537398	0.568051
6	0.124872258	0.177818	0.230764	0.253057	0.328296	0.350589	0.442548	0.487134	0.54008	0.562373	0.593026
7	0.14984671	0.202793	0.255739	0.278032	0.35327	0.375563	0.467522	0.512108	0.565054	0.587347	0.618
8	0.209700413	0.27321	0.336719	0.36346	0.45371	0.480451	0.590757	0.644238	0.707747	0.734488	0.771257
9	0.239657614	0.303167	0.366676	0.393417	0.483667	0.510408	0.620714	0.674195	0.737705	0.764445	0.801214
10	0.269614816	0.333124	0.396634	0.423374	0.513624	0.540365	0.650671	0.704152	0.767662	0.794403	0.831171
11	0.299572018	0.363081	0.426591	0.453332	0.543582	0.570322	0.680628	0.73411	0.797619	0.82436	0.861128
12	0.32952922	0.393039	0.456548	0.483289	0.573539	0.60028	0.710585	0.764067	0.827576	0.854317	0.891086
13	0.359486422	0.422996	0.486505	0.513246	0.603496	0.630237	0.740543	0.794024	0.857533	0.884274	0.921043
14	0.389443623	0.452953	0.516462	0.543203	0.633453	0.660194	0.7705	0.823981	0.887491	0.914231	0.951
15	0.528632023	0.608682	0.688732	0.722438	0.836193	0.869898	1.008933	1.076343	1.156393	1.190099	1.236443
16	0.566391453	0.646442	0.726492	0.760197	0.873952	0.907658	1.046692	1.114103	1.194153	1.227858	1.274203
17	0.604150883	0.684201	0.764251	0.797956	0.911712	0.945417	1.084451	1.151862	1.231912	1.265617	1.311962
18	0.641910314	0.72196	0.802011	0.835716	0.949471	0.983177	1.122211	1.189622	1.269672	1.303377	1.349722
19	0.679669744	0.75972	0.83977	0.873475	0.987231	1.020936	1.15997	1.227381	1.307431	1.341136	1.387481
20	0.717429174	0.797479	0.877529	0.911235	1.02499	1.058695	1.19773	1.26514	1.34519	1.378896	1.425241
21	0.755188604	0.835239	0.915289	0.948994	1.06275	1.096455	1.235489	1.3029	1.38295	1.416655	1.463
22	1.03335233	1.137672	1.241991	1.285915	1.434159	1.478083	1.659269	1.747117	1.851437	1.895361	1.955756

	Freshwater										
Day	**0**	**1140**	**2280**	**2760**	**5220**	**5700**	**7320**	**7800**	**8940**	**9420**	**10,080**
23	1.082559584	1.186879	1.291199	1.335123	1.483366	1.52729	1.708477	1.796325	1.900644	1.944568	2.004964
24	1.131766838	1.236086	1.340406	1.38433	1.532573	1.576497	1.757684	1.845532	1.949851	1.993775	2.054171
25	1.180974092	1.285294	1.389613	1.433537	1.581781	1.625705	1.806891	1.894739	1.999059	2.042983	2.103378
26	1.230181346	1.334501	1.43882	1.482744	1.630988	1.674912	1.856098	1.943946	2.048266	2.09219	2.152585
27	1.2793886	1.383708	1.488028	1.531952	1.680195	1.724119	1.905306	1.993154	2.097473	2.141397	2.201793
28	1.328595853	1.432915	1.537235	1.581159	1.729402	1.773326	1.954513	2.042361	2.14668	2.190604	2.251
29	1.838344307	1.977533	2.116723	2.175328	2.373124	2.431729	2.673479	2.790691	2.92988	2.988486	3.069069
30	1.903999461	2.043189	2.182378	2.240984	2.438779	2.497385	2.739134	2.856346	2.995535	3.054141	3.134724
31	1.969654615	2.108844	2.248033	2.306639	2.504434	2.56304	2.804789	2.922001	3.06119	3.119796	3.200379
32	2.035309769	2.174499	2.313688	2.372294	2.570089	2.628695	2.870444	2.987656	3.126845	3.185451	3.266035
33	2.100964923	2.240154	2.379343	2.437949	2.635744	2.69435	2.9361	3.053311	3.192501	3.251107	3.33169
34	2.166620076	2.305809	2.444998	2.503604	2.701399	2.760005	3.001755	3.118967	3.258156	3.316762	3.397345
35	2.23227523	2.371464	2.510653	2.569259	2.767054	2.82566	3.06741	3.184622	3.323811	3.382417	3.463
36	3.12125033	3.310309	3.499368	3.578972	3.847634	3.927238	4.255603	4.414811	4.60387	4.683473	4.792929
37	3.210428911	3.399488	3.588547	3.66815	3.936813	4.016417	4.344782	4.503989	4.693048	4.772652	4.882107
38	3.299607492	3.488666	3.677725	3.757329	4.025991	4.105595	4.433961	4.593168	4.782227	4.861831	4.971286
39	3.388786072	3.577845	3.766904	3.846507	4.11517	4.194774	4.523139	4.682347	4.871405	4.951009	5.060464
40	3.477964653	3.667023	3.856082	3.935686	4.204349	4.283952	4.612318	4.771525	4.960584	5.040188	5.149643
41	3.567143234	3.756202	3.945261	4.024865	4.293527	4.373131	4.701496	4.860704	5.049763	5.129366	5.238821
42	3.656321815	3.845381	4.034439	4.114043	4.382706	4.46231	4.790675	4.949882	5.138941	5.218545	5.328

Table 5.
Freshweight sensitivity analysis from freshwater.

A duration-based control using fuzzy logic system was developed to translate the output amplitude of the nutrient solution automation into a time dimension. This controls the duration of the solenoid valve in turning on and off for the NPK solution to flow. For the short duration, the solenoid valve should be on for 5 seconds to give 480 mL liquid concentration. For the long duration, the solenoid valve should be on for 10 seconds to give 660 mL liquid concentration. This process only occurs once a day as the system was set to capture image information on daily basis only. The artificial lighting system was tested at different photoperiods of 9, 12, and 16 hours, on layers 1, 2, and 3 of growth beds, respectively. Results of these are shown in **Table 5**. The phytopigments were measured for one sample of every layer. It was observed from the result that the optimal photoperiod is at 12 hours. Hence, the light setting of the system should be continuously on for 12 hours and off for the next 12-hour cycle.

Sensitivity analysis has proven to be a significant tool in determining the effectivity of models used in eco-systems. The response of the actuators for nutrient, conditioning, and irrigation control was measured on how effective they were on taking automation controls based on the changes on the input data. The analysis contains eight tables of information for the four phytopigments considered such as fresh weight, chlorophyll-A, chlorophyll-B, and vitamin C. The performance of the actuators was measured based on cumulative freshwater volume and cumulative NPK solution volume. The two actuator measurements were used since the time the crop was first planted to determine the changes of the four phytopigments that resulted to eight combinations of sensitivity analysis dataset. The dataset contains 42 samples representing every phytopigment value captured each day for the six-week cultivation. The samples were defined through what if analysis from 42 days of measuring the amount of NPK solution and freshwater added to the hydroponics unit each day from the fuzzy logic controller that operates once every 24 hours. This results to a 42-by-42 dataset on which the measured amount of liquid is added cumulatively for each sample. To simplify the representation, the 42 measured liquid substances added to the system were summarized into 11 samples that were proportionately parted to represent the whole dataset. Note that at day 42, the standard value for the lettuce crops at week 6 for harvest was accurately obtained. **Table 5** shows sample fresh weight sensitivity analysis from freshwater. When a total of 10,800 mL of freshwater was added on the water inflowing to the hydroponics growth bed, the freshweight obtained was 5.328 g. It is interesting to mention that the required standard value for the lettuce crops at week 6 for harvest is obtained accurately in all experiment results.

5. Conclusions and recommendations

The wireless sensor nodes for irrigation control, nutrient mixture automation, adaptive temperature maintenance, and lighting systems between the hydroponic chambers and the pond for aquaculture were effectively implemented for the automation and control of the adaptive management system. The water flow as controlled by the irrigation system successfully recirculates from the pond to the mixing tank and then flows through the hydroponics chamber back to the pond through the drain and conditioning tank. The nutrient mixture automation depends on the machine vision system data, deep learning models, and fuzzy logic controller to determine the amount of nutrient solutions to be added on the liquid concentration before it flows to the growth beds. The adaptive temperature maintenance as controlled by fuzzy logic maintains the standard temperature and humidity for optimal crop growth. Lighting the crops is dependent on an artificial real-time clock setup that has a photoperiod of 9, 12, and 16 hours, respectively, for each layer of the growth beds.

Data acquired from the wireless sensor nodes were utilized to determine the responses of the actuators. For the irrigation control, each of the tanks contains a pump or solenoid valves that are reliant to the water level and nutrient sensors. For the nutrient mixing tank, there are three water-level sensors that control the solenoid valve for the pond water to control if the inflow should be stopped while there is a continuous outflow of water to the hydroponics chamber. Mixing tanks are controlled to determine whether the tank should be filled with nutrient concentrations. The outflow, however, is not controlled by the irrigation control but by the nutrient mixture automation. The conditioning tank has the same irrigation control with the nutrient mixing tank. The drain irrigation, however, is only controlled by two water-level sensors that turn on the pump if the tank is almost full and turn off the pump as the tank almost empties. The nutrient mixture automation relies on data acquired from the vision systems. Maintaining the temperature suitable for lettuce growth is reliant on the temperature and humidity data acquired from five sensors across the chamber.

Transmission of data that are used for determining the actuators' response is done wirelessly to store them on a database and to present them on a cloud-based monitoring system through a common router node. Sending the data to the common router node is integrated with the use of the built-in WIFI module of the ESP32 through the wireless transmission program developed in each of the sensor nodes.

Machine vision acquires image data that are wirelessly sent to the cloud-based database and monitoring as the input for the crop growth optimization based on the phytochemical and phytopigment and fresh weight models. The models were trained using computational-based algorithms. Lettuce crop images underwent image processing techniques to obtain the data that are used as response variables to be predicted by the deep learning networks. Predictions made by the models from newly acquired data that were not yet seen or used by the model are integrated to a fuzzy logic controller to determine the duration of solenoid valve opening once per day. This controls the nutrient concentration added to the mixing tank per day to adaptively adjust the nutrients to be absorbed by the plants based on its yield from the previous day. The adaptive management system on the nutrient mixture automation both in the mixing and conditioning tank is using the models to automatically set the required nutrients needed by the cultivars in the hydroponics chamber and the aquaculture pond.

Sensitivity analysis was then used to determine whether the adaptive management system responds timely and accurately based on the input data from the wireless sensor networks and the machine vision systems. The analysis determines whether the amount of nutrients cumulatively added per day result in the needed concentration for the crops. According to the standard values from **Table 4**, the system was able to obtain the necessary amount of phytopigments per cultivation week as the nutrients were adjusted accordingly. During the harvesting stage, the lettuce crops, fresh weight, chlorophyll-A, chlorophyll-B, and vitamin C were at 5.328 g, 540.48 mg/L, 790.48 mg/L, and 790.48 mg/L, respectively, that were the optimal values to harvest the crops at its optimal yield.

For future works, it is recommended that an adaptive management system for different crops will be implemented in a single-smart aquaponics system. The concept underlies a way for switching an AMS that can produce an optimal yield for different cultivars that can be cultivated in a single system of hydroponics unit and aquaculture unit. This will address the limitation of study that solely focuses on lettuce crops. Other ecological and environmental factors aside from the four aspects covered in the study can be added to the AMS as well such as gas system emission control to determine the crops' growth effect as well to other surrounding cultivars that will be added in the system.

Author details

Elmer P. Dadios[1]*, Ryan Rhay Vicerra[1], Sandy Lauguico[2], Argel Bandala[2], Ronnie Concepcion II[1] and Edwin Sybingco[2]

1 Department of Manufacturing Engineering and Management, De La Salle University, Manila, Philippines

2 Department of Electronics and Computer Engineering, De La Salle University, Manila, Philippines

*Address all correspondence to: elmer.dadios@dlsu.edu.ph

References

[1] Kyalo Willy D, Muyanga M, Jayne T. Can economic and environmental benefits associated with agricultural intensification be sustained at high population densities? A farm level empirical analysis. Land Use Policy. 2019;**81**(May 2018):100-110. DOI: 10.1016/j.landusepol.2018.10.046

[2] Fujimoto S, Mizuno T, Ohnishi T, Shimizu C, Watanabe T. Relationship between population density and population movement in inhabitable lands. Evolutionary and Institutional Economics Review. 2017;**14**(1):117-130. DOI: 10.1007/s40844-016-0064-z

[3] Ricker-Gilbert J, Jumbe C, Chamberlin J. How does population density influence agricultural intensification and productivity? Evidence from Malawi. Food Policy. 2014;**48**:114-128. DOI: 10.1016/j.foodpol.2014.02.006

[4] Song J, Tong X, Wang L, Zhao C, Prishchepov AV. Monitoring finer-scale population density in urban functional zones: A remote sensing data fusion approach. Landscape and Urban Planning. 2019;**190**(January):103580. DOI: 10.1016/j.landurbplan.2019.05.011

[5] Huttunen S. Revisiting agricultural modernisation: Interconnected farming practices driving rural development at the farm level. Journal of Rural Studies. 2019;**71**(January):36-45. DOI: 10.1016/j.jrurstud.2019.09.004

[6] Abu Hatab A, Cavinato MER, Lindemer A, Lagerkvist CJ. Urban sprawl, food security and agricultural systems in developing countries: A systematic review of the literature. Cities. 2019;**94**(June):129-142. DOI: 10.1016/j.cities.2019.06.001

[7] Yue S, Munir IU, Hyder S, Nassani AA, Qazi Abro MM, Zaman K. Sustainable food production, forest biodiversity and mineral pricing: Interconnected global issues. Resources Policy. 2020;**65**(August 2019):101583. DOI: 10.1016/j.resourpol.2020.101583

[8] Tian X, Xu X. Urban agriculture and urban sustainable development. In: 2012 6th International Association for China Planning Conference, IACP 2012. Wuhan, China: IEEE; 2012. DOI: 10.1109/IACP.2012.6401979

[9] Chaudhry AR, Mishra VP. A comparative analysis of vertical agriculture systems in residential apartments. In: 2019 Advances in Science and Engineering Technology International Conferences, ASET 2019. Dubai, United Arab Emirates: IEEE; 2019. pp. 1-5. DOI: 10.1109/ICASET.2019.8714358

[10] Yazgac BG, Durmus H, Kirci M, Gunes EO, Karli HB. Petri nets based procedure of hardware/software codesign of an urban agriculture monitoring system. In: 2019 8th International Conference on Agro-Geoinformatics, Agro-Geoinformatics. Istanbul, Turkey: IEEE; 2019. DOI: 10.1109/Agro-Geoinformatics.2019.8820255

[11] Mcdougall R, Rader R, Kristiansen P. Urban agriculture could provide 15% of food supply to Sydney, Australia, under expanded land use scenarios. Land Use Policy. 2020;**94**(February 2019):104554. DOI: 10.1016/j.landusepol.2020.104554

[12] Chen J. Rapid urbanization in China: A real challenge to soil protection and food security. Catena. 2007;**69**(1):1-15. DOI: 10.1016/j.catena.2006.04.019

[13] Nuwansi KKT, Verma AK, Rathore G, Prakash C, Chandrakant MH, Prabhath GPWA. Utilization of phytoremediated aquaculture

wastewater for production of koi carp (Cyprinus carpio var. koi) and gotukola (Centella asiatica) in an aquaponics. Aquaculture. 2019;**507**(September 2018):361-369. DOI: 10.1016/j.aquaculture.2019.04.053

[14] Zaini A, Kurniawan A, Herdhiyanto AD. Internet of things for monitoring and controlling nutrient film technique (NFT) aquaponic. In: 2018 International Conference on Computer Engineering, Network and Intelligent Multimedia, CENIM 2018 - Proceeding. Surabaya, Indonesia: IEEE; 2018. pp. 167-171. DOI: 10.1109/CENIM.2018.8711304

[15] Yang T, Kim HJ. Nutrient management regime affects water quality, crop growth, and nitrogen use efficiency of aquaponic systems. Scientia Horticulturae. 2019;**256** (March):108619. DOI: 10.1016/j.scienta.2019.108619

[16] Mori J, Smith R. Transmission of waterborne fish and plant pathogens in aquaponics and their control with physical disinfection and filtration: A systematized review. Aquaculture. 2019;**504**(February):380-395. DOI: 10.1016/j.aquaculture.2019.02.009

[17] Yep B, Zheng Y. Aquaponic trends and challenges – A review. Journal of Cleaner Production. 2019;**228**: 1586-1599. DOI: 10.1016/j.jclepro.2019.04.290

[18] Gómez C, Currey CJ, Dickson RW, Kim HJ, Hernández R, Sabeh NC, et al. Controlled environment food production for urban agriculture. HortScience. 2019;**54**(9):1448-1458. DOI: 10.21273/HORTSCI14073-19

[19] Duston J. Assessing the Potential Environmental Impacts of Controlled Environment Agriculture in Detroit and the Future of This Industry Based on Local Food Trends [Doctoral dissertation]. Cambridge, MA, United States: Harvard Extension School; 2017. Available from: http://proxyiub.uits.iu.edu/login?url=https://search.ebscohost.com/login.aspx?direct=true&db=edshld&AN=edshld.1.33826456&site=eds-live&scope=site.

[20] Lakhiar IA, Gao J, Syed TN, Chandio FA. Modern plant cultivation technologies in agriculture under controlled environment : A review on aeroponics. Journal of Plant Interactions. 2018;**13**(1):338-352. DOI: 10.1080/17429145.2018.1472308

[21] Sánchez-escobar F, Coq-huelva D, Sanz-cañada J. Measurement of sustainable intensification by the integrated analysis of energy and economic flows: Case study of the olive-oil agricultural system of Estepa, Spain. Journal of Cleaner Production. 2018; **201**:463-470. DOI: 10.1016/j.jclepro.2018.07.294

[22] Jiao X, Zhang H, Ma W, Wang C, Li X, Zhang F. Science and Technology Backyard : A novel approach to empower smallholder farmers for sustainable intensification of agriculture in. Journal of Integrative Agriculture. 2019;**18**(8):1657-1666. DOI: 10.1016/S2095-3119(19)62592-X

[23] Sivakami S, Karthikeyan C. Expert Systems with Applications Evaluating the effectiveness of expert system for performing agricultural extension services in India. Expert Systems with Applications. 2009;**36**(6):9634-9636. DOI: 10.1016/j.eswa.2008.11.054

[24] Fan J, Fang L, Wu J, Guo Y, Dai Q. From brain science to artificial intelligence. Engineering. 2020;**6**(3): 248-252. DOI: 10.1016/j.eng.2019.11.012

[25] Saleem G, Akhtar M, Ahmed N, Qureshi WS. Automated analysis of visual leaf shape features for plant classification. Computers and Electronics in Agriculture. 2019;**157** (November 2018):270-280. DOI: 10.1016/j.compag.2018.12.038

[26] Dhingra G, Kumar V, Dutt H. A novel computer vision based neutrosophic approach for leaf disease identification and classification. Measurement. 2019;**135**:782-794. DOI: 10.1016/j.measurement.2018.12.027

[27] Tang D, Feng Y, Gong D, Hao W, Cui N. Evaluation of artificial intelligence models for actual crop evapotranspiration modeling in mulched and non-mulched maize croplands. Computers and Electronics in Agriculture. 2018;**152**(March):375-384. DOI: 10.1016/j.compag.2018.07.029

[28] Kyaw TY, Ng AK. Smart aquaponics system for urban farming. Energy Procedia. 2017;**143**:342-347. DOI: 10.1016/j.egypro.2017.12.694

[29] He S, Wang H, Bo H, Hu S. Studies on informational intervention strategies of the urban community agriculture. In: Proceedings - 2016 International Conference on Smart Grid and Electrical Automation, ICSGEA 2016. Zhangjiajie, China: IEEE; 2016. pp. 332-335. DOI: 10.1109/ICSGEA.2016.23

[30] Pölling B, Sroka W, Mergenthaler M. Success of urban farming's city-adjustments and business models—Findings from a survey among farmers in Ruhr Metropolis, Germany. Land Use Policy. 2017;**69**(January):372-385. DOI: 10.1016/j.landusepol.2017.09.034

[31] Armanda DT, Guinée JB, Tukker A. The second green revolution: Innovative urban agriculture's contribution to food security and sustainability – A review. Global Food Security. 2019;**22**(August 2018):13-24. DOI: 10.1016/j.gfs.2019.08.002

[32] Odame HS, Owuo JBO, Changeh JG, Otieno JO. ScienceDirect The role of technology in inclusive innovation of urban agriculture. Current Opinion in Environmental Sustainability. 2020;**43**:1-6. DOI: 10.1016/j.cosust.2019.12.007

[33] Magwaza ST, Magwaza LS, Odindo AO, Mditshwa A. Hydroponic technology as decentralised system for domestic wastewater treatment and vegetable production in urban agriculture: A review. Science of the Total Environment. 2020;**698**:134154. DOI: 10.1016/j.scitotenv.2019.134154

[34] Chen S. Industrial biosystems engineering and biorefinery systems. Chinese Journal of Biotechnology. 2008;**24**(6):940-945. DOI: 10.1016/S1872-2075(08)60044-8

[35] Nitisoravut R, Regmi R. Plant microbial fuel cells: A promising biosystems engineering. Renewable and Sustainable Energy Reviews. 2017;**76** (September 2016):81-89. DOI: 10.1016/j.rser.2017.03.064

[36] Mo C, Kim G, Kim MS, Lim J, Lee K, Lee WH, et al. On-line fresh-cut lettuce quality measurement system using hyperspectral imaging. Biosystems Engineering. 2017;**156**:38-50. DOI: 10.1016/j.biosystemseng.2017.01.005

[37] Weng S, Zhu W, Zhang X, Yuan H, Zheng L, Zhao J, et al. Recent advances in Raman technology with applications in agriculture, food and biosystems: A review. Artificial Intelligence in Agriculture. 2019;**3**:1-10. DOI: 10.1016/j.aiia.2019.11.001

[38] González García M, Fernández-López C, Bueno-Crespo A, Martínez-España R. Extreme learning machine-based prediction of uptake of pharmaceuticals in reclaimed water-irrigated lettuces in the Region of Murcia, Spain. Biosystems Engineering. 2019;**177**:78-89. DOI: 10.1016/j.biosystemseng.2018.09.006

[39] Norton T, Berckmans D. Engineering advances in Precision Livestock Farming. Biosystems Engineering. 2018;**173**:1-3. DOI: 10.1016/j.biosystemseng.2018.09.008

[40] Rojo-Gimeno C, van der Voort M, Niemi JK, Lauwers L, Kristensen AR, Wauters E. Assessment of the value of information of precision livestock farming: A conceptual framework. NJAS - Wageningen Journal of Life Sciences. 2019;**90–91**(November): 100311. DOI: 10.1016/j.njas.2019.100311

[41] Perakis K, Lampathaki F, Nikas K, Georgiou Y, Marko O, Maselyne J. CYBELE – Fostering precision agriculture & livestock farming through secure access to large-scale HPC enabled virtual industrial experimentation environments fostering scalable big data analytics. Computer Networks. 2020; **168**:107035. DOI: 10.1016/j.comnet.2019.107035

[42] Benis K, Ferrão P. Commercial farming within the urban built environment – Taking stock of an evolving fi eld in northern countries. Global Food Security. 2018;**17** (November 2017):30-37. DOI: 10.1016/j.gfs.2018.03.005

[43] Goodman W, Minner J. Will the urban agricultural revolution be vertical and soilless? A case study of controlled environment agriculture in New York City. Land Use Policy. 2019;**83**(June 2018):160-173. DOI: 10.1016/j.landusepol.2018.12.038

[44] Lefers RM, Srivatsa Bettahalli NM, Fedoroff NV, Ghaffour N, Davies PA, Nunes SP, et al. Hollow fibre membrane-based liquid desiccant humidity control for controlled environment agriculture. Biosystems Engineering. 2019;**183**:47-57. DOI: 10.1016/j.biosystemseng.2019.04.010

[45] Magsumbol JV, Baldovino RG, Valenzuela IC, Sybingco E, Dadios EP. An Automated Temperature Control System : A Fuzzy Logic Approach. In: 2018 IEEE 10th International Conference on Humanoid, Nanotechnology, Information Technology, Communication and Control, Environment and Management (HNICEM). Baguio, Philippines: IEEE; 2018. pp. 1-6. DOI: 10.1109/HNICEM.2018.8666239

[46] Viršilė A, Brazaitytė A, Vaštakaitė-Kairienė V, Miliauskienė J, Jankauskienė J, Novičkovas A, et al. The distinct impact of multi-color LED light on nitrate, amino acid, soluble sugar and organic acid contents in red and green leaf lettuce cultivated in controlled environment. Food Chemistry. 2020; **310**:125799. DOI: 10.1016/j.foodchem.2019.125799

[47] Harun AN, Ahmad R, Mohamed N. Plant growth optimization using variable intensity and Far Red LED treatment in indoor farming. In: 2015 International Conference on Smart Sensors and Application, ICSSA 2015. Harbin, China: IEEE; 2015. pp. 92-97. DOI: 10.1109/ICSSA.2015.7322517

[48] El-Nakhel C, Petropoulos SA, Pannico A, Kyriacou MC, Giordano M, Colla G, et al. The bioactive profile of lettuce produced in a closed soilless system as configured by combinatorial effects of genotype and macrocation supply composition. Food Chemistry. 2020;**309**:125713. DOI: 10.1016/j.foodchem.2019.125713

[49] Li C, Adhikari R, Yao Y, Miller AG, Kalbaugh K, Li D, et al. Measuring plant growth characteristics using smartphone based image analysis technique in controlled environment agriculture. Computers and Electronics in Agriculture. 2020;**168**(October 2019): 105123. DOI: 10.1016/j.compag.2019.105123

[50] Jung HY, Kim JK. Complete reutilisation of mixed mackerel and brown seaweed wastewater as a high-quality biofertiliser in open-flow lettuce hydroponics. Journal of Cleaner Production. 2020;**247**:119081. DOI: 10.1016/j.jclepro.2019.119081

[51] R. Conception, S. Lauguico, J. Alejandrino, A. Bandala, R, Vicerra, E. P. Dadios, J. Cuello, Adaptive fertigation system using hybrid vision-based lettuce phenotyping and fuzzy logic valve controller towards sustainable aquaphonics, Journal of Advanced Computational Intelligence and Intelligent Informatics, Vol. 25 No.5, pp. 610-617, 2021

[52] Batarseh FA, Ramamoorthy G, Dashora M, Yang R. Intelligent automation tools and software engines for managing federal agricultural data. In: Federal Data Science: Transforming Government and Agricultural Policy Using Artificial Intelligence. USA: Academic Press; 2018. DOI: 10.1016/B978-0-12-812443-7.00012-0

[53] Escarabajal-Henarejos D, Molina-Martínez JM, Fernández-Pacheco DG, García-Mateos G. Methodology for obtaining prediction models of the root depth of lettuce for its application in irrigation automation. Agricultural Water Management. 2015;**151**:167-173. DOI: 10.1016/j.agwat.2014.10.012

[54] Paraforos DS, Vassiliadis V, Kortenbruck D, Stamkopoulos K, Ziogas V, Sapounas AA, et al. Multi-level automation of farm management information systems. Computers and Electronics in Agriculture. 2017;**142** (November):504-514. DOI: 10.1016/j.compag.2017.11.022

[55] Palande V, Zaheer A, George K. Fully automated hydroponic system for indoor plant growth. Procedia Computer Science. 2018;**129**:482-488. DOI: 10.1016/j.procs.2018.03.028

[56] Kaburuan ER, Jayadi R, Harisno. A design of IoT-based monitoring system for intelligence indoor micro-climate horticulture farming in Indonesia. Procedia Computer Science. 2019;**157**: 459-464. DOI: 10.1016/j.procs.2019.09.001

[57] A. L. P. de Ocampo, E. P. Dadios, Integrated weed estimation and pest damage detection in Solanum melongena plantation via aerial vision-based proximal sensing, Philippine Journal of Science 150 (4): 677-688, August 2021

[58] Boursianis AD, Papadopoulou MS, Diamantoulakis P, Liopa-Tsakalidi A, Barouchas P, Salahas G, et al. Internet of things (iot) and agricultural unmanned aerial vehicles (UAVs) in smart farming: A comprehensive review. Internet of Things. 2020:100187. DOI: 10.1016/J.IOT.2020.100187 (in press)

[59] Hang L, Ullah I, Kim DH. A secure fish farm platform based on blockchain for agriculture data integrity. Computers and Electronics in Agriculture. 2020;**170** (December 2019):105251. DOI: 10.1016/j.compag.2020.105251

[60] dos Santos MJPL. Smart cities and urban areas—Aquaponics as innovative urban agriculture. Urban Forestry and Urban Greening. 2016;**20**:402-406. DOI: 10.1016/j.ufug.2016.10.004

[61] Karimanzira D, Rauschenbach T. Enhancing aquaponics management with IoT-based Predictive Analytics for efficient information utilization. Information Processing in Agriculture. 2019;**6**(3):375-385. DOI: 10.1016/j.inpa.2018.12.003

[62] Suhl J, Oppedijk B, Baganz D, Kloas W, Schmidt U, van Duijn B. Oxygen consumption in recirculating nutrient film technique in aquaponics. Scientia Horticulturae. 2019;**255**(May):281-291. DOI: 10.1016/j.scienta.2019.05.033

[63] Pérez-Urrestarazu L, Lobillo-Eguíbar J, Fernández-Cañero R, Fernández-Cabanás VM. Suitability and optimization of FAO's small-scale aquaponics systems for joint production of lettuce (Lactuca sativa) and fish (Carassius auratus). Aquacultural Engineering. 2019;**85**(February):

129-137. DOI: 10.1016/j.aquaeng.2019.04.001

[64] Paudel SR. Nitrogen transformation in engineered aquaponics with water celery (*Oenanthe javanica*) and koi carp (*Cyprinus carpio*): Effects of plant to fish biomass ratio. Aquaculture. 2020;**520**: 734971. DOI: 10.1016/j.aquaculture.2020.734971

[65] Witzel O, Wilm S, Karimanzira D, Baganz D. Controlling and regulation of integrated aquaponic production systems – An approach for a management execution system (MES). Information Processing in Agriculture. 2019;**6**(3):326-334. DOI: 10.1016/j.inpa.2019.03.007

[66] Mamatha MN, Namratha SN. Design and implementation of indoor farming using automated aquaponics. System. 2017;**2**(August):396-401

[67] Michalak D. Adapting to climate change and effective water management in Polish agriculture – At the level of government institutions and farms. Ecohydrology and Hydrobiology. 2020;**20**(1):134-141. DOI: 10.1016/j.ecohyd.2019.12.004

[68] Memarzadeh M, Boettiger C. Adaptive management of ecological systems under partial observability. Biological Conservation. 2018;**224**(May):9-15. DOI: 10.1016/j.biocon.2018.05.009

[69] Ramli MR, Daely PT, Kim D, Lee JM. IoT-based adaptive network mechanism for reliable smart farm system. Computers and Electronics in Agriculture. 2020;**170**(July 2019): 105287. DOI: 10.1016/j.compag.2020.105287

[70] Castaldi P. Adaptive Signal Processing Strategy for a Wind Farm System Fault Accommodation. IFAC-PapersOnLine. n.d.;**51**(24):52-59. DOI: 10.1016/j.ifacol.2018.09.528

[71] Paas W, Groot JCJ. Creating adaptive farm typologies using Naive Bayesian classification. Information Processing in Agriculture. 2017;**4**(3): 220-227. DOI: 10.1016/j.inpa.2017.05.005

[72] Patrício DI, Rieder R. Computer vision and artificial intelligence in precision agriculture for grain crops: A systematic review. Computers and Electronics in Agriculture. 2018;**153**(April):69-81. DOI: 10.1016/j.compag.2018.08.001

[73] Jha K, Doshi A, Patel P, Shah M. A comprehensive review on automation in agriculture using artificial intelligence. Artificial Intelligence in Agriculture. 2019b;**2**:1-12. DOI: 10.1016/j.aiia.2019.05.004

[74] Pantazi XE, Moshou D, Bochtis D. Artificial intelligence in agriculture. In: Intelligent Data Mining and Fusion Systems in Agriculture. 2020. pp. 17-101. DOI: 10.1016/b978-0-12-814391-9.00002-9

[75] Partel V, Kakarla SC, Ampatzidis Y. Development and evaluation of a low-cost and smart technology for precision weed management utilizing artificial intelligence. Computers and Electronics in Agriculture. 2019;**157**(November 2018):339-350. DOI: 10.1016/j.compag.2018.12.048

[76] Bu F, Wang X. A smart agriculture IoT system based on deep reinforcement learning. Future Generation Computer Systems. 2019;**99**:500-507. DOI: 10.1016/j.future.2019.04.041

[77] Domingues DS, Takahashi HW, Camara CAP, Nixdorf SL. Automated system developed to control pH and concentration of nutrient solution evaluated in hydroponic lettuce production. Computers and Electronics in Agriculture. 2012;**84**:53-61. DOI: 10.1016/j.compag.2012.02.006

Chapter 2

A Review of BIM-Based Automated Code Compliance Checking: A Meta-Analysis Research

Murat Aydın

Abstract

This study aims to discuss the articles which are only available in electronic academic databases and only written in English in the subject of BIM and ACCC. Statistical results about the articles are obtained by the meta-analysis. In this study, meta-analysis method was selected as the method, and the general situation of the articles is presented based on statistical analysis data in the AEC industry. For meta-analysis, adapting from previous studies, a meta-analysis template consisting of four main categories is created and each category is subdivided into sub-categories. As a result of the study conducted in the electronic academic databases of the ITU (Istanbul Technical University) library website, there are 168 studies, including 131 articles and 37 proceedings until 31.12.2020. Only articles are analyzed in this study, and proceedings are not included. A literature review is carried out in the publications on the subject of ACCC through BIM by the ITU library website's academic databases to determine in which research trend is, in which areas are concentrated, and which areas are available. The study gives detailed information about the articles on ACCC subject area in the AEC industry in sub-categories when making general evaluation in categories.

Keywords: automated code compliance checking, building information modeling, common graphs, industry foundation classes, meta-analysis

1. Introduction

Building Information Modeling (BIM) is a simulation prototyping technology. The definition of the BIM according to the US National BIM Standard is "BIM is the digital representation of the physical and functional characteristics of a building project." The concept of BIM emerged in the 1970s, is discussed by academicians in various publications. The importance of BIM technology was increased gradually since the 2000s. Various BIM software has been developed by computer-aided design application providers such as Autodesk®, Graphisoft®, Bentley®, etc. In 1997, a new data standard Industry Foundation Classes (IFC), was established by Industry Alliance for Interoperability (IAI) [1, 2]. IFC is independent of any software, and it is a standard object-based data standard developed in express modeling language [3, 4]. IFC is supported by leading BIM-based software. Thus, it is widely accepted that BIM and IFC data standard will significantly improve and facilitate cooperation in the design processes.

A high level of interoperability and cooperation is essential to enhance expertise and automation in the construction industry. The required data must be represented correctly according to the types, characteristics, and names to build a building. A validated IFC building model is a vital prerequisite for executing many automated tasks such as building performance analysis and Automated Code Compliance Checking (ACCC) [5]. ACCC, according to building regulations, is an important task that must be handled carefully during the whole design process. Manually building regulation checking, a traditional method is a repetitive, time-consuming, and error-prone process for architects, engineers, or public authorities [6–8]. That is why BIM's effective automated code compliance checking is considered a promising domain in Architecture, Engineering, and Construction (AEC) industry.

ACCC studies are generally directed at standardizing and automating problems encountered in building regulation control [9]. Although ACCC studies were conducted in different countries, it is also understood that these countries show similarities to the problems set by the researchers. It is seen that most of the scientific studies aimed at local administration, which is in the most crucial position in the implementation and control of the building production process under the legislation, are limited to the subject area of process improvement and geographical information system [10–13]. As a result of the literature research being studied on the subject area, it is seen that there are no general conclusions about the ACCC domain; and it is needed to combine the results of the research conducted in different countries. On the other hand, emerging BIM technology and ACCC have become critical issues to evaluate building performance in design and building permit processes automatically. It is also a fact that it is too late to start working on ACCC in a country where the construction sector is so active. Some local administrations are observed to investigate the problems and state of the art manual code checking, and the ACCC possibilities are examined; a BIM-based model has been designed and coded. In this article, the literature review results in the first stage of the research project are given. This study is addressing the interaction between BIM and ACCC.

Articles on the subject of BIM and ACCC, which are accessible in the electronic academic databases, are analyzed in the paper. In this study, where the findings are presented in pivot tables and graphs, meta-analysis method was preferred. Tables and graphs are developed with the help of Microsoft Excel® and Graph Commons website. A meta-analysis template is transformed by adapting some previous meta-analysis studies. The template consists of four main categories be about Data, Content, Form & input-output relationship, and Purpose & outcome relationship and each category is subdivided into sub-categories. Findings are expressed with the help of pivot tables and graphs as the results of the meta-analysis.

2. Background of the ACCC studies

ACCC is a rule-based method that provides simultaneous control of building elements and building regulations, considering the building elements' size and characteristics and their associated regulations and codes [14–16]. In ACCC, structural elements and components are checked for compliance with the relevant regulation's rules and conditions, and checking reports are produced [9]. The domain of automated rules, code, and regulation compliance checking has been interested in many researchers and practitioners over the years. It started in 1960 as a subfield of Artificial Intelligence (AI) and linguistics to investigate human language's automatic creation and comprehension. The first computable rule development study dates back to the 1960s in the AEC industry. The American Institute of Steel Construction (AISC) specifications had been made using a decision table, and it has been formalized in these years. The research conducted by Fenves in 1966, using the regulations' decision tables, is shown as the first scientific study in this area in literature [17]. In his research, Fenves created different decision tables with regulation rules and conditions under which these rules may apply [18]. In 1973, a project was undertaken to restructure the AISC specification. In this project, a theoretical model was presented in which a four-layer structure can explain the knowledge of regulation. This regulation model was used as the SASE modeling methodology (Standard, Analysis, Synthesis, and Expression) software developed by Fenves and his team in 1987 [19]. This software was developed as a tool to provide the organization of regulation rules, decision tables, information networks, and classification systems. Towards the end of the 1980s, studies began about the development of building ACCC systems.

When it came to the 1990s, building models and rule checking methods have been developed, but effective computable code systems have more recently begun to emerge. The first study on ACCC has been CORENET in Singapore since 1995. As of 2000, researches and studies about building BIM-based design evaluation have increased. The latest and current ACCC Systems are Construction and Real Estate NETwork (CORENET), FORNAX, Solibri Model Checker (SMC), Jotne EDMmodelChecker & The Express Data Manager (EDM), Statsbygg, International Code Council (ICC) & SMARTcodes, General Services Administration (GSA) & Design Assessment Tool (DAT), Korean Research Studies, DesignCheck, LicA, ACCBEP, GTPPM, Natural Language Processing (NLP), and Artificial Intelligence (AI) [20]. The present studies have shown the possibility of various rule generation approaches and applicability to ACCC by many countries. The Republic of Korea, Norway, Portugal, The United States of America, Australia, The Republic of Singapore have given importance to BIM-based ACCC studies towards increasing the quality of design. Private or public institutions have financed most of these studies. Some of the studies deal with building regulation as a whole, and some only involve fire, elevator, water system, safety, building envelope, installation, circulation, parking, etc. regulations.

3. Background of meta-analysis

The literature review allows us to learn the state of the recent or current literature. Review articles can cover a wide range of subject matter at various levels of completeness and comprehensiveness. These are based on analyses of literature that may include research findings. There are various types of literature reviews, including meta-analysis/systematic review, scoping review, rapid review, umbrella review, and systematized review [21].

Quantitative methods of combining study results were first described in the early 1930s and grew in interest, especially in health, in the 1970s. Glass gave a "Meta-analysis" name for the kind of his research in 1976 [22]. Applications of meta-analysis to accumulated research literature showed that research findings were not nearly as conflicting as had been thought and that valuable and sound general conclusions could be drawn from existing research [23]. Meta-analysis began to evolve with Doll and Peto's intensive work in Oxford in the 1980s [24]. Wasserman, Hedges, Olkin, and Petitti defined the meta-analysis's statistical methods [25, 26]. Greenland described the statistical methods for meta-analysis of non-experimental studies [27]. At present, many discoveries and advances in cumulative knowledge are being made not by those who do primary research studies but by those who use meta-analysis to discover the latent meaning of existing research literature [28, 29].

The classification is of great importance to determine any domain's analysis, how the research trend is focused, in which areas it is concentrated, and in which areas there are spaces [30]. By analyzing the studies carried out within the scope, it is expected to give information about ACCC studies. A detailed literature review is carried out for the studies within the subject's scope [29]. Many articles, proceedings, papers, theses, book chapters, research reports, published or unpublished sources are accessed through scanning [28].

Meta-analysis is a method of combining the results of multiple studies independent of each other; and statistical analysis of research findings [28]. It combines the results of studies done in different places, times, and centers on the same subject, and it makes a general conclusion about the related topic [31].

Meta analysis using quantitative methods to synthesize and summarize results is a systematic review. It can be completely objective in evaluating research findings. This advantage makes it different and preferable from other reviews. Hence, in this paper, the meta-analysis method analyzes ACCC studies in the BIM domain.

- The research questions for each article are as follows:
- What is the basic info (keywords, authors, year, institution, country, etc.)?
- What is the content?
- What are the form, input, and output?
- What are the purpose and outcome?

3.1 The objective and limitations of the meta-analysis study

It is necessary to examine previous research results to develop a conceptual model for ACCC. This study aims to find the research gap and organize the existing information to form the conceptual model's basis to be developed. Articles published between 01.01.1988 and 31.12.2020 are included in the review. Only leading international electronic academic databases are used within the scope of the meta-analysis study (e.g., Emerald, John Wiley & Sons, ICONDA CIB Library, Science Direct, ASCE, Taylor & Francis, Web of Science, ITcon, Scopus, Engineering Village, and Google Scholar) [32]. "Building Information Modeling," "BIM," "IFC," "Industry Foundation Classes," "Code Checking," "Automated Code Compliance Checking," "Code Compliance Checking," "Building Code" keywords are used in Title, Abstract and Keywords sections of advanced search screen of the databases. The search areas are limited to:

- Architectural design,
- Automation,
- Building codes,
- Engineering,
- Environmental science,
- Construction industry, and
- Construction building technology.

As a result of the search in the academic databases in ITU Library, 168 studies, including 131 articles and 37 proceedings, are found (**Figure 1**). To achieve a confident academic standard, 131 articles that are peer-reviewed are included in the analysis. The only

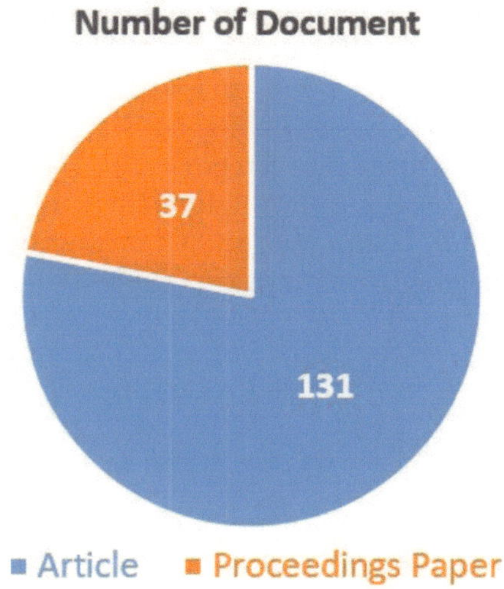

Figure 1.
Numerical distribution of document type.

peer-reviewed articles are analyzed in the meta-analysis study. Papers in proceedings that generally include the progress of preliminary research are not included in the scope.

3.2 Template of meta-analysis study

Meta-analysis template allows the articles analyzed within the scope to be categorized under the headings and subheadings discussed. The meta-analysis template, which consists of four main categories be about Data, Content, Form & input-output relationship, and Purpose & outcome relationship, shown in **Table 1**,

Category	Sub-category
1. Data	1.1. Article type
	1.2. Keyword
	1.3. Author
	1.4. Institution
	1.5. Country
	1.6. Source
	1.7. Publisher
	1.8. Year
	1.9. Database
	1.10. Area
2. Content	2.1. Subject
	2.2. Study level
	2.3. Process
3. Form & input-output relationship	3.1. Method
	3.2. Data type
	3.3. Contribution
4. Purpose & outcome relationship	4.1. Problem
	4.2. Tool
	4.3. Result

Table 1.
The meta-analysis template.

was prepared with the help of the frameworks of similar studies, and each category was subdivided into subcategories to provide detailed information [28, 33].

The Data Category consists of 10 sub-categories: e.g., Title, Keyword, Author, Institution, Country, Source, Publisher, Year, Database, and Area. The content category consists of three sub-categories: e.g., Subject, Study level, and process. The form & input-output relationship category consists of three sub-categories: e.g., Method, Data Type, and Contribution. The purpose & outcome relationship category consists of three sub-categories: e.g., Problem, Tool, and Result.

4. Methodology and analysis

131 articles were transformed into a comprehensive table with Microsoft Excel®, which shows the findings on the pivot tables and graphs. Bar charts were used for the "Data" category. In other categories, interactive map network graphs via the Graph Commons website were used to express sub-categories' relationships [34]. Graph Commons is a collaborative platform for mapping, analyzing, and publishing data-networks. It empowers researchers, people, and organizations to transform their data into interactive maps and untangle complex relations. In Graph Common, it needs at least two fundamental values for a map networks graph. It establishes the results with a Name typed in From Type, Edge To, and Type To columns in an Excel format file. In an interactive map networks graph, the ones with high weight (the arrow and the text) become more dominant than the others, which shows their importance.

4.1 Data

"Data" category consists of 10 sub-categories be about Title, Keyword, Author, Institution, Country, Source, Publisher, Year, Database, and Area of an article. The results of the analysis are given under the related headings. Since the results are more remarkable in number in sub-categories, the results less than three are grouped under the "Others."

4.1.1 Article type

The "Type" of articles used for the meta-analysis is examined in the Article Type sub-category (**Figure 2**), which shows articles' distribution by article types. 98 of the 131 articles are published directly in journals, while the remaining 33 are published as selected papers at congresses.

4.1.2 Keywords

The keywords are listed in the "Keyword" sub-category. The distribution of the keywords is shown in **Figure 2**. Articles with less than three keywords are categorized into the "Others" sub-category to limit the number of keywords. The most frequently used keywords are "Building information modeling (BIM)" (58 pcs), "Industry foundation classes (IFC)" (21 pcs), "Building information model" (13 pcs), and "Building codes" (13 pcs). Subsequently; Rule checking, Interoperability, Code checking, Computer-aided design, Automated code compliance checking, Semantic systems, Natural Language Processing, Model checking, Building design, Automated compliance checking, Prevention through design, Open BIM, Governance, Compliance checking, Code compliance checking, Building regulations, Automation, Automated information extraction, Automated code checking, and AEC industry are the most common keywords.

4.1.3 Author

The "Author" sub-category includes the authors' names and numbers. The distribution of authors is shown in **Figure 2**. It is seen that the number of articles in which 2 authors are the highest, 7 and 8 authors is the lowest. In general, the articles are prepared by large groups. This preparation shows the subject's multidisciplinary face, and the research studies should be supported by all the disciplines involved in the construction activities.

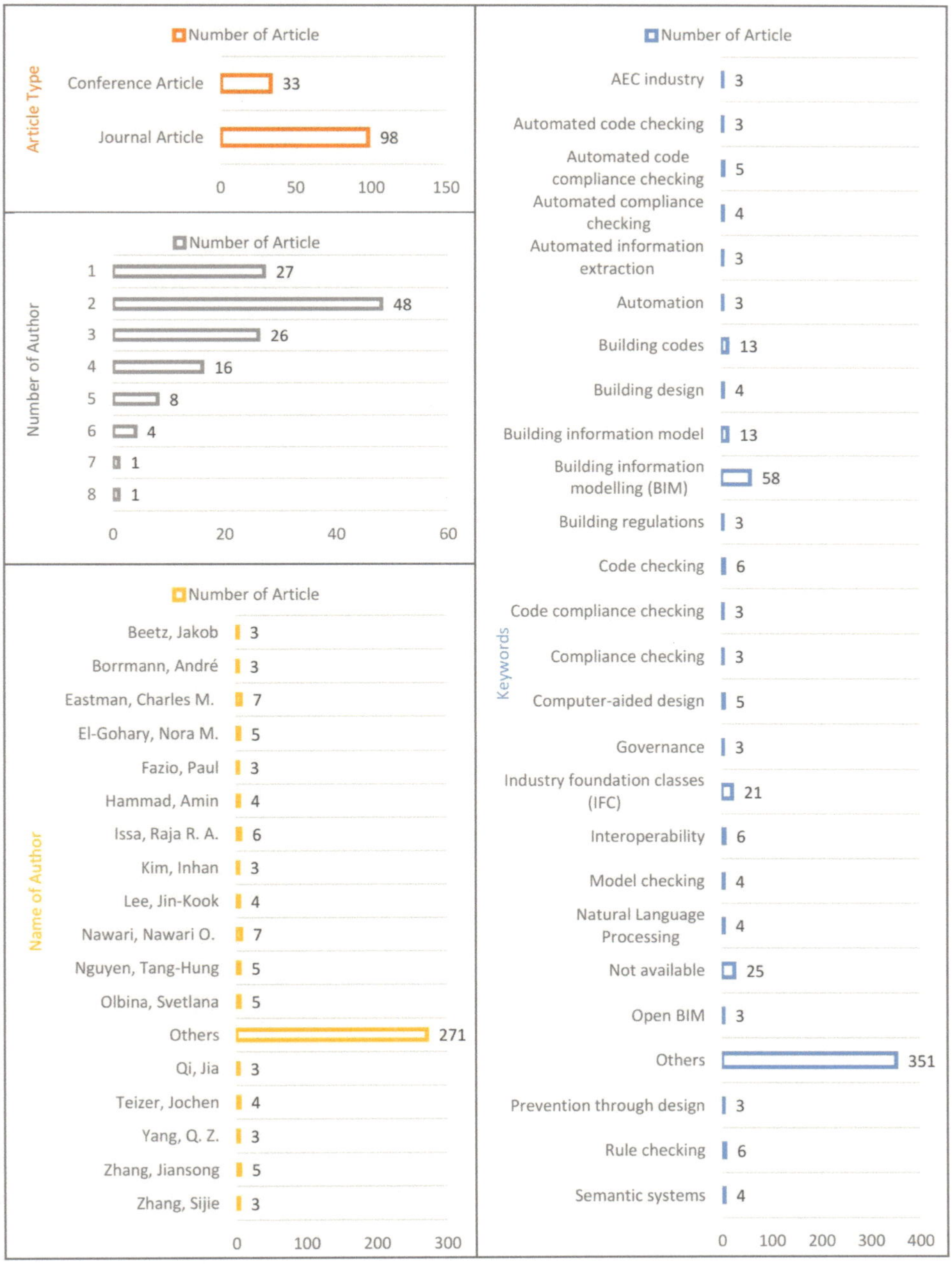

Figure 2.
Distribution of article type, keyword, and author (data) according to the number of articles.

4.1.4 Institution

The authors' institutions are grouped in the "Institution" sub-category. **Figure 3** shows the distribution of articles by institution type. Most of the articles prepared in Universities (111 pcs) as the findings of the research projects. Then University-Private Sector (8 pcs), Private Sector (4 pcs), University-Research Center-Private Sector (4 pcs) follows the universities. When University-Research Center-Private Sector-Public Sector associations collaboratively work, the first authors of the

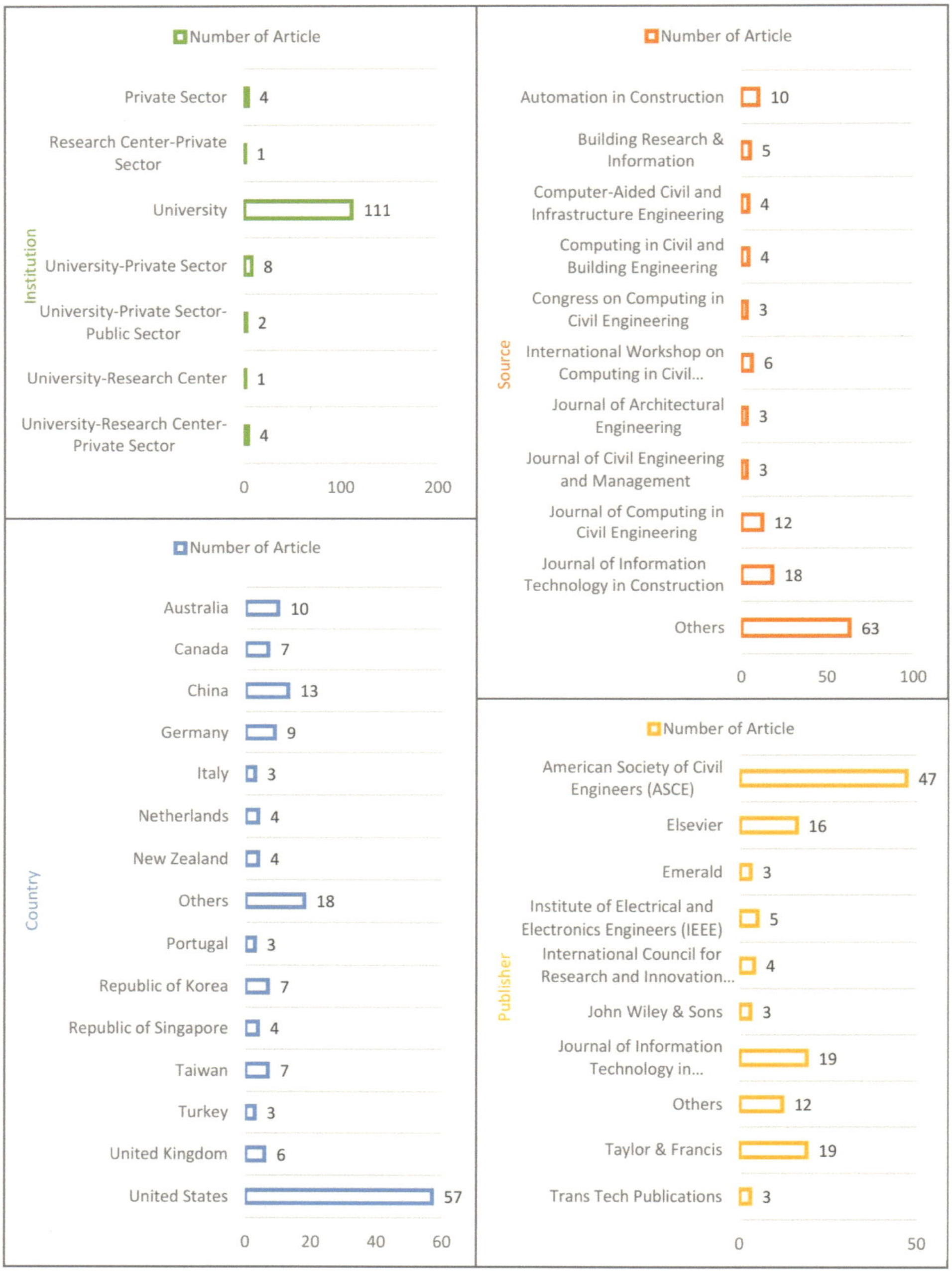

Figure 3.
Distribution of institution, country, source, and publisher (data) according to the number of articles.

articles are mostly academicians. Accordingly, it can be seen that the construction sector and university cooperation is currently limited in the subject area of BIM and ACCC, even though cooperation is a very urgent need. Collaborative research should be supported to reduce public institutions' workload, increase efficiency, and minimize possible human errors.

4.1.5 Country

The countries of the institutions are grouped in the "Country" sub-category (**Figure 3**). Countries with fewer than three are grouped under "Others." Most of the articles are prepared in the USA (57 pcs). China (13 pcs), Australia (10 pcs), and Germany (9 pcs) follow the USA.

4.1.6 Source

The Source sub-category shows the source of the article. The distribution of articles by sources can be seen in **Figure 3**. Most of the articles are published in the Journal of Information Technology in Construction (18 pcs), Journal of Computing in Civil Engineering (12 pcs), and Automation in Construction (10 pcs). Subsequently, Workshop on Computing in Civil Engineering (6 pcs), Building Research & Information (5 pcs), Computing in Civil and Building Engineering (4 pcs), Computer-Aided Civil and Infrastructure Engineering (4 pcs), Journal of Civil Engineering and Management (3 pcs), Journal of Architectural Engineering (3 pcs) and Congress on Computing in Civil Engineering (3 pcs) sources are available.

4.1.7 Publisher

The "Publisher" sub-category shows the publishers of the articles. The distribution of articles by publishers can be seen in **Figure 4**. Most of the articles are published by the American Society of Civil Engineers (ASCE) (47 pcs). The others are respectively Journal of Information Technology in Construction (ITcon) (19 pcs), Taylor & Francis (19 pcs), Elsevier (16 pcs), Institute of Electrical and Electronics Engineers (IEEE) (5 pcs), International Council for Research and Innovation in Building and Construction (CIB) (4 pcs), Emerald (3 pcs), John Wiley & Sons (3 pcs) and Trans Tech Publications (3 pcs).

4.1.8 Year

The year in which the articles were published is grouped in the "Year" sub-category. The distribution of articles by years can be seen in **Figure 4**. The earliest study on BIM, IFC, and ACCC was published in 1988, and the number of articles has increased steadily between 1995 and 2011 [35]. There has been an increase in the number of articles as of 2012. It peeks in 2014 (26 pcs) and 2016 (22 pcs). BIM and ACCC have gained importance after 2011.

4.1.9 Database

The electronic academic databases are analyzed in the "Database" sub-category. The distribution of the databases used in the research is shown in **Figure 4**. Apart from these, Google Scholar, ICONDA CIB Library, and ITcon academic databases are included in the meta-analysis study. Most of the articles are obtained from

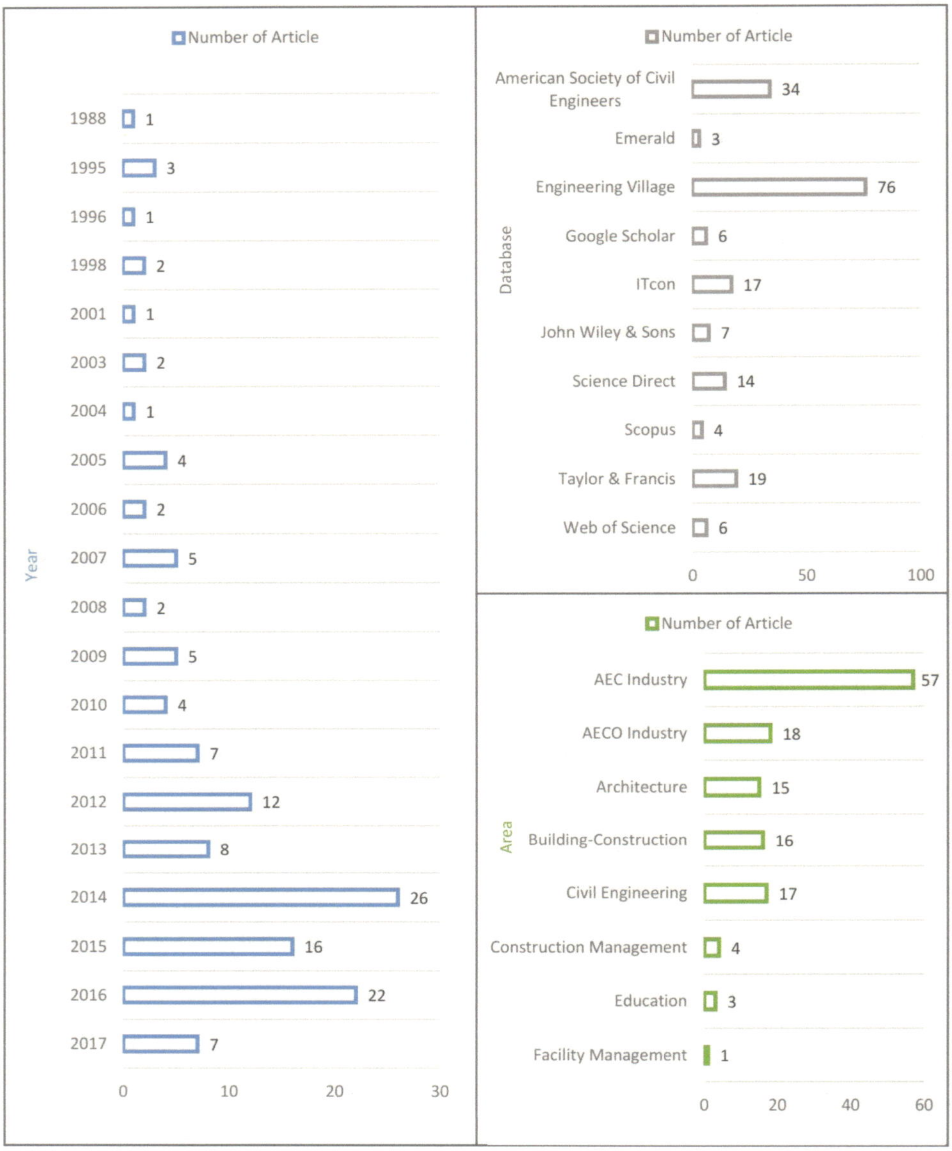

Figure 4.
Distribution of year, database, and area (data) according to the number of articles.

Engineering Village (76 pcs), and the most minor articles are accessed from Emerald (3 pcs). American Society of Civil Engineers (34 pcs), Taylor & Francis (19 pcs), ITcon (17 pcs), Science Direct (14 pcs) follows them respectively.

4.1.10 Area

The "Area" sub-category identifies the professional areas in which the articles are relevant in the construction sector. Although some of the articles are evaluated for a specific area such as Architecture, Civil Engineering, Education, there are also some articles in the AEC Industry and AECO Industry, including many professional areas. As seen in **Figure 4**, most of the articles are performed in Architecture, Engineering and Construction Industry (AEC Industry) (57 pcs). Architecture, Engineering, Construction and Operation Industry (AECO Industry) (18 pcs), Civil Engineering

(17 pcs), Building-Construction (16 pcs), Architecture (15 pcs), Construction Management (4 pcs), Education (3 pcs) follows it respectively.

4.2 Content

"Content" category, which is the second category of meta-analysis' framework, consists of 3 sub-categories be about Subject, Study Level, and Process. The relationship between these sub-categories is shown in **Figure 5**, a map networks graph prepared by Common Graph. The Starting Point (Red Dots) shows the "Subject" sub-category, the Edge (Colorful Arrows) shows the "Study Level" sub-category, and the End Point (Blue Dots) shows the "Process" sub-category. As a result of the analysis, the distribution of the number of articles obtained in each sub-category is given in brackets in the map networks graph, and bar charts are automatically prepared according to this graph. The results of the analysis are given in detail under the related sub-category. According to the map networks graph in **Figure 5**, the articles are few, and they are not dominant, which are done in the same Subject, Study, and Process levels.

4.2.1 Subject

The topics of the articles are grouped under the "Subject" sub-category. In the Subject sub-category, the perspective of BIM, IFC, and ACCC concepts is analyzed. As shown in **Figure 5**, most articles are based on Automated Code Compliance Control (35 pcs). Then the articles on BIM Application (19 pcs), Building Code (17 pcs), BIM-based Rule Checking (10 pcs), BIM-based Model Checking (10 pcs), and Interoperability (9 pcs) subjects are prominent. As a result of this grouping, some articles have specific topics such as BIM-based Automatic Programming (6 pcs), Computer-Aided Design (5 pcs), Architectural Design (5 pcs), Structural Design (4 pcs), Construction Project Management (4 pcs), Visualization (2 pcs), Process Control (2 pcs), BIM Education (2 pcs) and Construction Contracts (1 pcs).

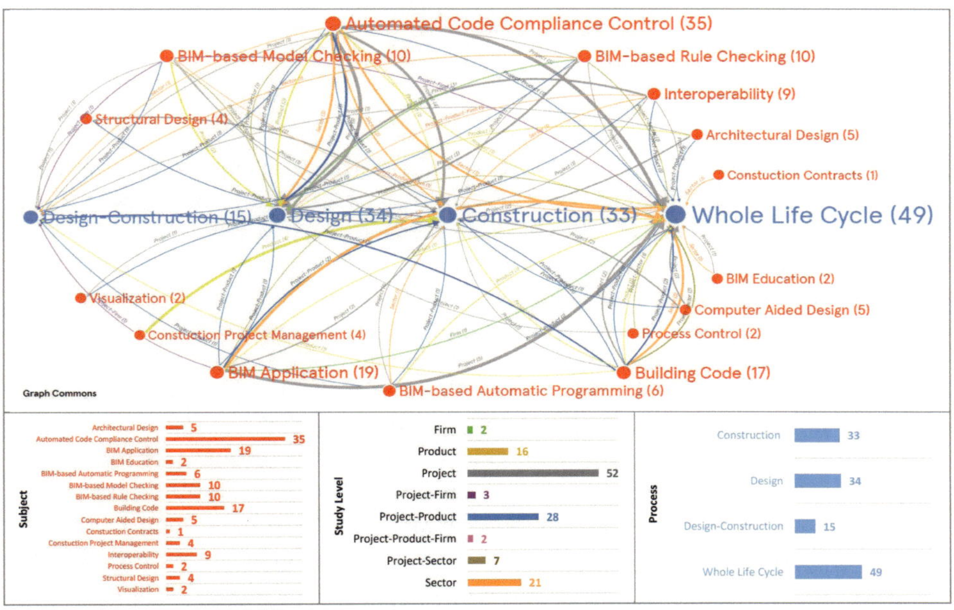

Figure 5.
Distribution of subject, process, and study level (content) according to the number of articles.

4.2.2 Study level

Articles are grouped at four basic levels in the "Study Level" sub-category. These levels are namely Project, Sector, Firm, and Product. Most of the articles are dealt with the Project level. Therefore, Project-Product, Project-Sector, Project-Firm, Project-Product-Firm levels are also created. As seen in **Figure 5**, most articles are discussed at the Project level (52 pcs). Project-Product (28 pcs), Sector (21 pcs), Product (16 pcs), Project-Sector (7 pcs), Project-Firm (3 pcs), Project-Product-Firm (2 pcs), and Firm (2 pcs) levels follow it respectively.

4.2.3 Process

The "Process" sub-category determines the relevant phase of the articles in the building production process. Planning, design, bid, construction, operation, and usage phases are grouped under Whole Life Cycle's title. As shown in **Figure 5**, the articles are primarily directed to the Whole Life Cycle (49 pcs). Other articles, respectively, are prepared for Design (34 pcs), Construction (33 pcs), and Design-Construction (15 pcs) processes.

4.3 Form & input-output relationship

Form & input-output relationship category consists of 3 sub-categories be about Method, Data Type, and Contribution. Form & input-output relationship is shown in **Figure 6**. The Starting Point (Red Dots) shows the "Method" sub-category, the Edge (Colorful Arrows) shows the "Data Type" sub-category, and the End Point (Blue Dots) shows the "Contribution" sub-category. The distribution of the number of articles obtained in each sub-category is given in brackets in the map networks graph, and bar charts are automatically created according to this graph. The results of the analysis are given in detail under the related sub-category. When we look at the map networks chart in general, it is seen that many articles (34 pcs) prefer the Case Study method, use Quantitative data type, and present Model Creation contribution.

4.3.1 Method

As shown in **Figure 6**, the Case Study (59 pcs) is the most common method used in the articles. Practical (34 pcs), Theoretical (20 pcs), Survey (11 pcs), and Interview (7 pcs) methods are also the other methods used in articles. The Case Study method is generally used in studies related to BIM-based residences or office buildings and unique case studies such as the validity of BIM and GIS integration, IFC construction models, the envelope design of the outer wall of a BIM-based building, and so on. Studies such as the development of BIM-based computable regulation rules and control, open-source IFC verification tool, and BIM-model control in building design are evaluated as Practical.

The other studies, such as literature review and BIM-based automated code compliance checking classification, are treated as Theoretical. The questionnaires conducted with the real estate sector participants, cost consultants, contractors, students, etc., on the adoption of BIM technology are evaluated as Surveys. The interview method is used in some studies with stakeholders of the construction sector such as expert project participants, designers, construction site staff and workers, project stakeholders, etc.

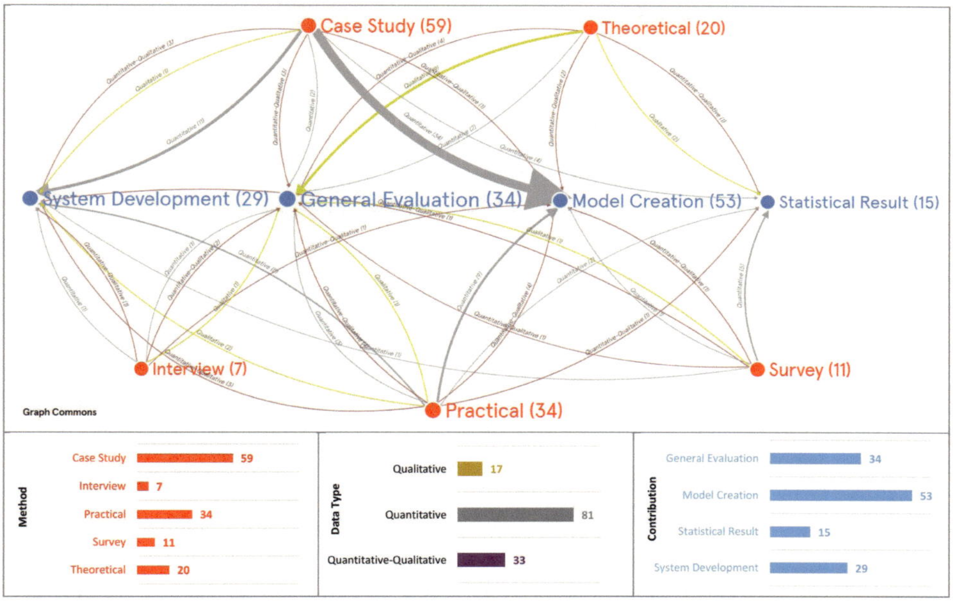

Figure 6.
Distribution of method, data type, and contribution (form & input-output relationship) according to the number of articles.

4.3.2 Data type

The type of data used in the articles is grouped in the "Data Type" sub-category. As seen in **Figure 6**, Most of the studies are based on Quantitative (81 pcs) data. The Quantitative data type is mainly seen in the studies which are utilized the case study method. It is used in the case studies for general and specific problems, such as implementation limitations of building code compliance checking, lack of interoperability in BIM applications, development of the automatic control and evaluation process for BIM data, results of BIM information acquisition difficulties, and control of green building design. Qualitative data obtained from the results of theoretical studies such as the failure to adopt BIM technology, the limitations of BIM in the construction sector, deficiencies of automated code compliance checking in the design process, and the limitations of existing building regulatory compliance control problems grouped. Quantitative-Qualitative data types are primarily obtained from articles where practical, theoretical, and case analysis methods are applied.

4.3.3 Contribution

The contribution of the article to the subject area is discussed in the "Contribution" sub-category. As shown in **Figure 6**, Model Creation (53 pcs), General Evaluation (34 pcs), System Development (29 pcs), and Statistical Findings (15 pcs) are the most known contributions of the studies, respectively. Model Creation is generally directed to the design of a BIM model. According to the BIM model, general evaluation is related to automated code compliance checking software, applications, and new trends. System Development focuses on the development of a BIM-based automated code compliance checking system. Statistical Result is based on the statistical results of the BIM-based model and code compliance checking application.

4.4 Purpose & outcome relationship

Purpose & Outcome Relationship category, the last category, consists of 3 sub-categories: Problem, Tool, and Result. The relationship between these sub-categories is shown in **Figure 7**. Starting Point (Red Dots) shows the Problem sub-category, the Edge (Colorful Arrows) shows the Tool sub-category, and the End Point (Blue Dots) shows the Result sub-category. As a result of the analysis, the distribution of the number of articles obtained in each sub-category is given in brackets in the map networks graph; and bar charts are automatically created according to this graph. The results of the analysis are given in detail under the related sub-category. In **Figure 7**, it is seen that the articles which are resulted from a Development for Requirement problem with the Software Development tool are dominant (20 pcs). The articles that resulted as a Proposal for Requirement problem with the Software Usage tool are also noteworthy (14 pcs).

4.4.1 Problem

The "Problem" sub-category evaluates the factors that led to the emergence of articles. As shown in **Figure 7**, Requirement is reached in most articles (60 pcs). The others are Lack (26 pcs), Result (12 pcs), Limitation (10 pcs), Insufficiency (7 pcs), Examination (6 pcs), Identification (5 pcs), and Divergence (5 pcs) problems. When we check the articles' reasons, the necessity of a BIM-based automated code compliance checking and national automated code compliance checking software problem comes to the fore. Furthermore, the importance of interoperability with practical BIM for computable building regulation rules is emphasized in many articles. Other problems result from BIM implementation difficulties in automated code compliance checking and the limitation of BIM implementation in the construction sector.

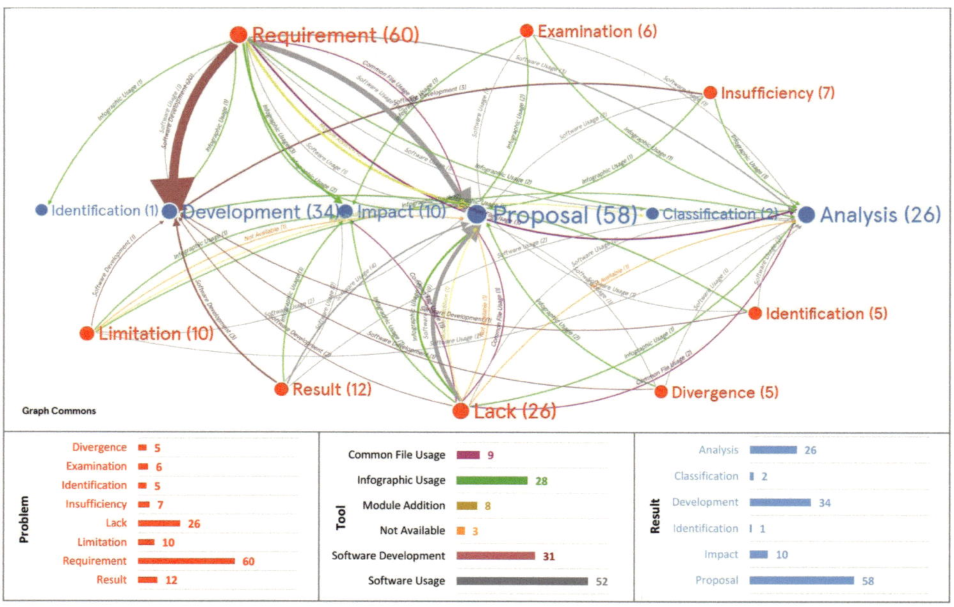

Figure 7.
Distribution of problem, tool, and result (purpose & outcome relationship) according to the number of articles.

4.4.2 Tool

"Tool" sub-category discusses the tools used to reach the goal in articles. As shown in **Figure 7**, Software Usage (52 pcs) is reached in most of the articles. The other tools are Software Development (31 pcs), Infographic Usage (28 pcs), Common File Usage (9 pcs), and Module Addition (8 pcs).

There are three articles in which tool information is not available. National (Lica, ACCBEP, GTPPM) software and International (SMC, Statsbygg, Fornax, Corenet, ePlanCheck, DesignCheck, SmartCodes, Jotne EDMmodelChecker) software are preferred as a Software Usage tool in the articles. Software Development tool concludes as a development for a requirement problem in most articles. Common File Usage tool transfers standard data files between software such as an IFC-BIM data exchange, a BIM and GIS integration, and a BIM and QR-code. Information (data sources, statistical information, contents, etc.) and visual expression (visual elements, graphics, images, etc.) are combined in a meaningful way with an Infographic Usage tool to create a graphical and visual representation. Module Addition tool gives additional features to existing software or systems.

4.4.3 Result

The results of each article are determined in the Result sub-category. As shown in **Figure 7**, a Proposal (58 pcs) is reached in most of the articles. Also; Development (34 pcs), Analysis (26 pcs), Impact (10 pcs), Classification (2 pcs), and Identification (1 pcs) results are obtained from the articles. The results of the articles emphasize the importance of automated code compliance checking. Several articles propose a new automated code compliance checking system. Except for proposals, the other results are the development of BIM-based automated code compliance checking system and software, the impact of interoperability analysis and automated code compliance checking system, the impact the classification of building regulation rules, and the identification of trends in BIM teaching.

5. Conclusions

This study carries out a literature review by analyzing the articles on ACCC through BIM by the ITU library website. It eliminates the preliminary study of BIM and ACCC. It gave the previous studies' general situation, research trends, and developed or proposed software features. Also, it creates a substructure for developing a BIM-based automated code compliance checking software in Turkey. The studies are investigated within the scope of BIM and ACCC subjects. The literature review covers the period between 1988 and 2020. As a result of the literature review, there are 168 publications, including 131 articles and 37 proceedings. This study only analyzes English articles except for proceedings. It evaluates the results by the meta-analysis method. The meta-analysis analyzes 131 articles by four main categories (Data, Content, Form & Input Relationship, Purpose & Outcome Relationship) and sub-categories within each category.

Of the 131 articles, 98 are journal articles, and 33 are conference articles. BIM, IFC, ACCC, and Building Codes are the keywords commonly used in articles. According to the keywords, articles include many professional areas such as architecture, construction, operation, computer science, etc. So, this situation shows that ACCC is related to different disciplines. It should be supported by all the disciplines involved in the construction sector. Most of the articles are written by academicians in universities. Most of the articles are carried out in the USA and published mainly by the American Society of Civil Engineers (ASCE) publisher by obtaining from the Engineering Village database.

The first ACCC article was published in 1988, and the number of articles has increased steadily between 1995 and 2011. After 2011, the increased number of articles shows the importance of BIM and ACCC. Although most articles are carried out in many professional fields, such as the AEC Industry and AECO Industry, some work in a specific area such as Architecture, Civil Engineering, Education. The articles impact the interoperability between ACCC and BIM Application. Most of them are dealt with the project level, associated with sector, firm, and product. They are primarily directed to the whole life cycle process. They generally apply to BIM-based housing or office buildings by using the case study method.

If we generally look at the articles' problems, the requirement of BIM-based automated code compliance checking software comes to the fore. They emphasize the importance of ACCC, and they often propose a new ACCC system. Lica, Statsbygg, SMC, Fornax, Corenet, ePlanCheck, DesignCheck, etc., have been developed for solving different problems. These systems help to standardize and automate the problems of building regulation compliance checking. The emerging problems are similar in other countries such as Turkey by analyzing them. As a result, this study creates a substructure for developing a BIM-based automated code compliance checking software for solving the problems in building regulation compliance checking faced by Turkish municipalities.

Conflict of interest

The author declares no conflict of interest.

Notes/thanks/other declarations

The author received no specific funding for this study.

Author details

Murat Aydın
Faculty of Architecture, Department of Architecture, Istanbul Technical University, Istanbul, Turkey

*Address all correspondence to: aydinmurat12@itu.edu.tr

References

[1] IAI. Release 1.0 IFC Model Architecture [Internet]. 1997. Available from: http://www.buildingsmart-tech.org/ifc/ [Accessed: January 1, 2018]

[2] Eastman CM. Building Product Models: Computer Environments, Supporting Design and Construction. London: CRC Press; 2018. p. 424. DOI: 10.1201/9781315138671

[3] ISO. Product Data Representation and Exchange Description Methods: The EXPRESS Language Reference Manual [Internet]. 1997. Available from: https://www.iso.org/standard/38047.html [Accessed: January 1, 2018]

[4] buildingSMART. buildingSMART International [Internet]. 2020. Available from: https://www.buildingsmart.org [Accessed: January 1, 2020]

[5] Eastman C. Automated assessment of early concept designs. Architectural Design. 2009;**79**(2): 52-57. DOI: 10.1002/ad.851

[6] Ding L, Drogemuller R, Rosenman M, Marchant D, Gero J. Automating code checking for building designs—DesignCheck. In: Clients Driving Innovation: Moving Ideas into Practice. Wollongong, Australia: Cooperative Research Centre (CRC) for Construction Innovation; 2006. pp. 1-16. https://ro.uow.edu.au/cgi/viewcontent.cgi?article=7773&context=engpapers

[7] Greenwood D, Lockley S, Malsane S, Matthews J. Automated compliance checking using building information models. In: Proceedings of the Construction, Building and Real Estate Research Conference. Paris, France: International Council for Research and Innovation in Building and Construction; 2010. pp. 363-371

[8] Aydın M, Yaman H. Domain knowledge representation languages and methods for building regulations. In: Communications in Computer and Information Science. Cham: Springer; 2020. pp. 101-121. DOI: 10.1007%2F978-3-030-42852-5_9

[9] Nawari NO. Automating codes conformance. Journal of Architectural Engineering. 2012 Dec;**18**(4):315-323. DOI: 10.1061/(ASCE)AE.1943-5568.0000049

[10] Demir R. Yapı ruhsatı ve yapı kullanma izni. Yerel Yönetim ve Denetim Dergisi. 2012;**17**(4):6-15

[11] Demir R. Planlı alanlar imar yönetmeliği hükümleri çerçevesinde yapı ruhsatı ve yapı kullanma izni. Yerel Yönetim ve Denetim Dergisi. 2017;**22**(10):29-42

[12] Reis YK. Belediyelerde imar ve yapı ruhsatı süreçlerinin etkinliğinin arttırılması. Planlama Dergisi. 2014;**24**(2):55-63. DOI: 10.5505/planlama.2014.83792

[13] Güler T, Coşgun N. Yapı üretim sürecinde belediyelerin rolü. Çağdaş Yerel Yönetimler. 2011;**20**(2):53-71. https://www.irbnet.de/daten/iconda/CIB20057.pdf

[14] Aydın M, Yaman H. An overview of building information modelling (bim) based building code compliance checking literatüre. Journal of Design+Theory. 2018;**14**(25):59-77. DOI: 10.14744/tasarimkuram.2018.25744

[15] Aydın M, Yaman H. A literature review of automated code compliance checking concept. Journal of Design+Theory. 2020;**16**(29):79-97. DOI: 10.14744/tasarimkuram.2020.86158

[16] Aydın M. Building information modeling based automated building regulation compliance checking asp.net web software. Intelligent Automation &

Soft Computing. 2021;**28**(1):11-25. DOI: 10.32604/iasc.2021.015065

[17] Fenves SJ. Tabular decision logic for structural design. Journal of the Structural Division. 1966;**92**(6):473-490. DOI: 10.1061/JSDEAG.0001567

[18] Fenves SJ, Gaylord EH, Goel SK. Decision Table Formulation of the 1969 AISC Specification. Urbana, IL: University of Illinois Engineering Experiment Station, College of Engineering, University of Illinois at Urbana-Champaign; 1969

[19] Fenves SJ, Wright RN, Stahl FI, Reed KA. Introduction to sase: Standards analysis, synthesis, and expression. Editorial National Bureau of Standards. 1987:473-490

[20] Nawari NO. SmartCodes and bim. Structures Congress. Reston, VA: American Society of Civil Engineers; 2013. pp. 928-937. DOI: 10.1061/9780784412848.082

[21] Grant MJ, Booth A. A typology of reviews: An analysis of 14 review types and associated methodologies. Health Information & Libraries Journal. 2009;**26**(2):91-108. DOI: 10.1111/j.1471-1842.2009.00848.x

[22] Glass GV. Primary, secondary, and meta-analysis of research. Educational Researcher. 1976;**5**(10):3-8. DOI: 10.3102/0013189X005010003

[23] Schmidt FL, Hunter JE. The validity and utility of selection methods in personnel psychology: Practical and theoretical implications of 85 years of research findings. Psychological Bulletin. 1998;**124**(2):262-274. DOI: 10.1037/0033-2909.124.2.262

[24] Doll R, Peto R. The causes of cancer: Quantitative estimates of avoidable risks of cancer in the United States today. JNCI: Journal of the National Cancer Institute. 1981;**66**(6):1192-1308. DOI: 10.1093/jnci/66.6.1192

[25] Petitti DB. Meta-Analysis, Decision Analysis, and Cost-Effectiveness Analysis: Methods for Quantitative Synthesis in Medicine. Oxford, UK: Oxford University Press; 2000. p. 320

[26] Wasserman S, Hedges LV, Olkin I. Statistical methods for meta-analysis. Journal of Educational Statistics. 1988;**13**(1):75-78. DOI: 10.2307/1164953

[27] Greenland S. Quantitative methods in the review of epidemiologic literature. Epidemiologic Reviews. 1987;**9**(1):1-30. DOI: 10.1093/oxfordjournals.epirev.a036298

[28] Betts M, Lansley P. Construction management and economics: A review of the first ten years. Construction Management and Economics. 1993 Jul;**11**(4):221-245. DOI: 10.1080/01446199300000024

[29] Hunter JE, Schmidt FL. Methods of Meta-Analysis. Thousand Oaks, CA: Sage Publication; 2004. p. 581

[30] Zhao X. A scientometric review of global bim research: Analysis and visualization. Automation in Construction. 2017;**80**:37-47. DOI: 10.1016/j.autcon.2017.04.002

[31] Volk R, Stengel J, Schultmann F. Building information modeling (bim) for existing buildings—Literature review and future needs. Automation in Construction. 2014;**38**:109-127. DOI: 10.1016/j.autcon.2013.10.023

[32] ITU Library. Istanbul Technical University Library Home/Databases [Internet]. 2020. Available from: http://kutuphane.itu.edu.tr/en/research/databases#harf-A [Accessed: December 31, 2020]

[33] Ilter T, Dikbas A, Ilter D. An analysis of drivers and barriers of construction innovation. In: Proceedings of 5th International

Conference on Innovation in Architecture, Engineering and Construction (AEC). Antalya, Turkey: Civil Engineering Department, Middle East Technical University & Centre for Innovative and Collaborative Engineering, Loughborough University; 23-25 June 2008. pp. 1-13. https://www.lboro.ac.uk/microsites/cice/aec2008/AEC2008-56.pdf

[34] Graph Commons. Graph Commons Map Networks Homepage [Internet]. 2018. Available from: https://graphcommons.com [Accessed: January 1, 2018]

[35] Arlani AG, Rakhra AS. Building code assessment framework. Construction Management and Economics. 1988;**6**(2):117-131. DOI: 10.1080/01446198800000011

Chapter 3

A Heuristically Generated Metric Approach to the Solution of Chase Problem

İhsan Ömür Bucak

Abstract

In this work, heuristic, hyper-heuristic, and metaheuristic approaches are reviewed. Distance metrics are also examined to solve the "puzzle problems by searching" in AI. A viewpoint is brought by introducing the so-called Heuristically Generated Angular Metric Approach (HAMA) through the explanation of the metrics world. Distance metrics are applied to "cat and mouse" problem where cat and mouse makes smart moves relative to each other and therefore makes more appropriate decisions. The design is built around Fuzzy logic control to determine route finding between the pursuer and prey. As the puzzle size increases, the effect of HAMA can be distinguished more clearly in terms of computation time towards a solution. Hence, mouse will gain more time in perceiving the incoming danger, thus increasing the percentage of evading the danger. 'Caught and escape percentages vs. number of cats' for three distance metrics have been created and the results evaluated comparatively. Given three termination criteria, it is never inconsistent to define two different objective functions: either the cat travels the distance to catch the mouse, or the mouse increases the percentage of escape from the cat.

Keywords: hyper-heuristics, metaheuristics, heuristically generated distance-metric, chase problem, fuzzy logic control

1. Introduction

Heuristics methods are usually employed to solve the AI search problems; however, in current approaches, heuristic algorithms are not always working. Indeed, a problem which is invalid for a heuristic approach can give successful results in an algorithmic approach. Heuristic approach, unlike algorithmic methods, does not show the exact path to reach the goal. The deficiency in question is not due to the heuristic methods but is related to the field of problem itself as well as the intuitiveness of the algorithms [1]. The features of the problems are also important in terms of heuristic programming. Solution accuracy in well-formed problems can be demonstrated by algorithmic approach. Theorem proofs can be given as an example to this kind of heuristic problems [1].

Time and memory limitation is concerned in resolving the well-formed problems by algorithmic approach on computers. For example, even though "Magic square" or "The Eight Queens" problems, at first glance, carry some sort of algorithmic nature, examination of all the situation yields combinatorial growth. Despite the fact that there are 362,880 states under discussion for a 3×3 sized magic square problem, the state space of the 5×5 sized problem grows so large that it cannot be searched thoroughly ($1, 5x10^{25}$ states). Once the problem state space grows larger, heuristic approach aims to search the solution in real-time. In light of these facts, the definition of intuitiveness according to Feldman and Feigenbaum, is given as follows [2]: "Intuitiveness once the state space of the problem becomes too large, is the usage of any rule, strategy, trick, simplification, and the other factors". Therefore, when the problem contains complexity, intuitiveness plays an important role to find the path to the solution [1].

The importance of heuristic metrics lies in guaranteeing the closest optimal solution in a short time in problems having variables with huge values. In toy problems such as an 8-puzzle, heuristic metrics which do the search, selection and optimization in a very short period of time can be seen as very critical. Optimization calculations of subatomic particles, and the use of heuristics in route finding, automation and city planning are some major applications which increase its importance. Nevertheless, the effects of investigation and analysis of solving behaviors of heuristic functions through puzzle problems can be observed easily as the puzzle size is increased. Thus, the better investigation or review possibility is provided.

This paper is organized as follows: The first section is an introduction to the research area and it describes the problem. The second section reviews heuristics, hyper-heuristics, and metaheuristics applied to solve AI problems from a broad literature perspective, including the relationships between them, while the third section introduces heuristically generated metric approach, and lays down its principles. The fourth section discusses a chasing problem between pursuers and pray. The design and development are built around fuzzy logic control to determine the paths of the pursuer and prey. The simulation results are reported and discussed in the fifth section. The last section summarizes the work with a conclusion.

2. A review of heuristics, hyper-heuristics, and metaheuristics with applications to solve AI problems

Burke et al. in their research have aimed to both seek the common goal of automating the design of heuristic metrics to solve computationally hard search problems and generalize search methodologies through the use of their introduced metric concept in solving the target problem [3]. They have classified the heuristics as "disposable" and "reusable" analogous to "on-line" and "off-line" learning, respectively. "On-line" learning heuristics learn a heuristic method while solving a given instance of a problem whereas "off-line" learning heuristics learn the method from a set of training instances to make generalization in the case of unseen examples. An example to on-line learning approach can be given as the use of reinforcement for heuristic selection, the use of hyper-heuristics as high-level search strategies and genetic algorithms in a search space of heuristics [4, 5]. Similarly, learning classifier systems, case-based reasoning, and genetic programming can be given as the examples to off-line learning [6–8]. Both heuristic approaches have their own advantages. For example, on-line learning methods or heuristics may provide a favorable structure to the search space; therefore, it can be more effective searching in the space of heuristics rather than in the space of problems directly. The other advantage may arise as an alternative solution against not having a set of related instances to train the heuristic methods off-line in newly encountered problems. A reusable or off-line method will always have an advantage of increasing the speed of solving new instances of problems [3].

As an example to the class of on-line learning algorithms, to improve the performance of Traveling Salesman Problem (TSP), a wide range of geometric heuristics have been investigated. The prime objective in terms of searching for a useful metric in TSP is a tour or a solution quality. Tour quality is assumed "good" when a valid tour is found even though only small percentage of output tours are valid tours (≈15% for random 10-city instances in the Euclidian metric). A valid tour and minimized cost (i.e., minimum tour length) are the major ones that affect the solution quality most out of the resulting tours. Usually, the incidence of valid tours are sought to be a part of any measure of success for performance [9].

A broad range of metrics are used in machine learning to evaluate the performance of metrics. Although there are many different metrics used in machine learning, a general theory is lacking to characterize the behavior of these metrics. For example, determining which metric can or should be used in a certain application or why a particular metric is a good metric is not clear. As some metrics behave differently from some other metrics, determining characteristics of these metrics can not be made precisely. Flach in their study has aimed to build a general theory and present a formal analysis in order to characterize the behavior of machine learning metrics such that dependence on these aspects become more precise and adequate. For instance, some metrics, by nature, do not depend on the class distribution in order to determine the amount of profit incurred for correctly classified examples, or misclassification cost distribution in order to determine the amount of cost incurred. The amount of cost or profit incurred is used to define the expected yield of a model. Some metrics such as precision, information gain, weighted relative accuracy have been used to build models in machine learning. Some others such as accuracy, F-measure, or area under ROC curve have been used to evaluate models on a test set. They have also proposed the use of true and false positive rates to evaluate the performance or the quality of models. Metric has been defined in terms of the counts of these true and false positive rates in a contingency table (a.k.a confusion matrix). Contingency table must have been chosen because it allows the metric to eliminate or exclude model complexity. It is also more suited to tabulate true and false positive rates as sufficient statistics for characterizing the performance of a classifier in any target context, and for evaluating the quality of a model. The metric also considers *skew ratio* as an additional parameter to indicate a trade-off between true and false positive rates to determine the direction in which improvements are to be found. Obviously, what a metric measures is no different from what an expected skew ratio is [10].

Another research aims to discover criteria that are responsible for a good performance of rule learning heuristics, theoretically and empirically. Rule learning heuristics aims to control a trade-off between consistency and coverage and therefore aims to determine optimizing parameters. The trade-off between consistency and coverage can only be explained through rules. The objective of this work is to discover criteria that are responsible for a good performance of rule learning heuristics, theoretically and empirically, and to understand the properties of rule learning heuristics such that those properties exhibit desirable performance for a large variety of datasets [11].

The term "metaheuristics" was first coined by Glover [12] and was used to explain a higher level strategy to modify other heuristics toward solutions generated beyond the search for local optimality [13]. Furthermore, a metaheuristics can be used to obtain quality solutions to the challenging optimization problems in a reasonable amount of time through the use of randomized local search algorithms under a certain tradeoff. Nearly, all of the metaheuristic algorithms show a tendency to work suitably for global optimization. However, metaheuristic algorithms do not guarantee to reach the optimal solutions, and do not work all the time either. There have been two main approaches to calculate the robustness of a particular solution in metaheuristic optimization. One is resampling and another one is the reuse of neighborhood solutions [14]. The first one is reliable but expensive whereas the latter one is unreliable but cheap. Mirjalili et al. have proposed a metric called confidence measure for metaheuristic optimization algorithms which aims to increase the reliability by an effective calculation of the confidence level for each solution during the optimization process. The authors have also proposed new confidence-based operations for robust metaheuristics and used these operators to design a confidence-based robust optimization algorithm which is applicable to different metaheuristics [14].

Early work was introduced on the approaches to the automatic heuristic generation during the period of late 70s and early 80s for less constrained subproblems (i.e., auxiliary problems) which were also called relaxed models [15–18]. These approaches were lacking systematic means to produce the problems and the models, and therefore they were mostly proved to be computationally expensive and inefficient [19]. Passino et al. have introduced a metric space approach to specify the "heuristic function" which is often difficult for the A^* algorithm. It was shown how to specify an admissible and monotonic heuristic function for a wide class of problem domains, which is, in general, too difficult to find one. The objective was also to reduce computational complexity or to obtain a huge computational savings in addition to the introduction of a new class of "good" heuristic functions which are admissible and monotone [20].

Rosenfeld et al. have focused on adaptive weight based heuristic approaches for dynamic and effective robot coordination to meet perceived environmental conditions such as coordination and spatial conflicts within the robot group. The authors have further stated that their robot coordination using interference metrics minimized interference and thus achieved high productivity [21].

Metric optimization problems have mostly been analyzed probabilistically as based on Euclidian instances. Nothing much has been done on the side of Non-Euclidian instances. Therefore, this has motivated Bringman et al. with a study of random metric instances for optimization problems which were obtained as based on a complete graph whose edges are assigned to random weights independently. The length of a shortest path between any two connecting nodes was specified as "distance". Then, the authors have proved structured properties of the random metric instances obtained as above. Their objective was to build good clusters. They have further used these results to analyze the approximation ratios of heuristics to match large-scale optimization problems such as TSP [22].

Akinyemi has investigated how to improve the play performance and game strategy of Ayo game through the enhancement of the knowledge of minimax search technique by using a refinement-based heuristic method [23].

Sosa-Ascencio et al. have proposed a variable solution heuristics generated by a grammar-based genetic programming framework to solve constraint satisfaction problems (CSPs) which is related to artificial intelligence. This approach is also called the newly generated grammar-based hyper-heuristic and therefore categorized under heuristic generation methodology, and is distinguished from heuristic selection methodology in that the former generates new heuristics from constituents of available heuristics whereas the latter chooses or selects available heuristics [24].

Another heuristic selection design for a grammar-based hyper-heuristic model has been applied this time to solve the two dimensional bin packing problem consisting of irregular pieces and regular objects. The objective of designing such a model was to select heuristics to determine the piece to be packed and the object in which the piece is located [25].

Dokeroglu and Cosar have proposed a novel multistart hyper-heuristic algorithm on the grid to solve the quadratic assignment problem (QAP) which is an NP-hard combinatorial optimization problem. The QAP is defined as the problem of assigning facilities to locations where each location has varying installation costs. The goal is to find an allocation where the total cost belongs to the installation and the transportation of the required amount of materials between the facilities is minimized. The proposed algorithm has put to use hyper-heuristics to find the best solution for an optimization problem by controlling and combining the strengths of several heuristics [26].

Wu et al. have proposed an evolutionary hyper-heuristic to solve the software project scheduling problem (SPSP). In a software project, an NP-hard combinatorial optimization problem, the SPSP assigns employees to tasks where the completion time, that is, the project duration, and the cost, that is, the total amount of salaries paid are to be minimized. The objective of this work is to find most suitable search operators for the type of the problem instance considered [27].

Lozano et al., in their approach, have used a parameterized schema of metaheuristics in which each metaheuristic or hybridized basic metaheuristics is represented by the values of a set consisting of numerical parameters. Their proposed schema aids the accomplishment of a metaheuristic selection or metaheuristics combination through the selection of the parameter values within the schema. The hyper-heuristic search of metaheuristic parameters in the metaheuristic space is automated except for the initial set-up of the hyper-heuristic parameters by the user. They finally decide to use a shared-memory parameterized schema both at the hyper-heuristic and the metaheuristic level, resulting in four level parallelism to reduce the solution time with high computation cost [28].

Another work explores a generation hyper-heuristic that automatically builds a selection hyper-heuristic using a machine learning algorithm called Time-Delay Neural Network (TDNN) used to extract hidden patterns within the collected data in the form a classifier, that is, an 'apprentice' hyper-heuristics, which is then used to solve the 'unseen' problem instances. The influence of extending and enriching the information collected from the expert and fed into TDNN is explored on the behavior of the generated apprentice hyper-heuristic [29].

In another work, the performance of a hyper-heuristic in terms of the quality and size of the heuristic pool is investigated. The objective is to produce a compact subset of effective heuristics from the unnecessary large pool that can decrease the performance of adaptive approaches. A new variant of iterated local search hyper-heuristics was also proposed, which incorporates dynamic multi-armed bandits. Both the heuristic pool selection method and the hyper-heuristic variant were successfully tested on two complex optimization problems: course timetabling and vehicle routing [30].

Whether sophisticated learning mechanisms are always necessary for hyper-heuristics to perform well has also been analyzed. For a benchmark function, Lissovoi et al. have proved that the Generalized Random Gradient Hyper-heuristics can learn to adapt the neighborhood size of Randomized Local Search to optimality during the run. They have also proved that the performance of the hyper-heuristics improves as the number of low-level local search heuristics to choose from increases [31].

Genetic programming has also been used as an offline hyper-heuristic to automatically evolve probability distributions, and hence to automatically generate mutation operators in an evolutionary programming as opposed to human designed existing operators [32].

Schlünz et al., in their work, have claimed that the first application of a multiobjective hyperheuristic, which is an evolutionary-based technique incorporating multiple sub-algorithms simultaneously, is applied to the multi-objective in-core fuel management (MICFMO) optimization problem. The hyperheuristic is able to raise the level of generality at which MICFMO may be performed, and it is capable of yielding improved quality in optimization results (compared to the preferred metaheuristics) [33].

A general-purpose selection hyper-heuristic search framework designed for the grouping has been extended to pair up various heuristic/operator selection and move acceptance methods for an NP-hard combinatorial optimization grouping problem of data clustering. The performances of various selection hyper-heuristics are compared using a set of benchmark instances which vary in terms of the number of items, groups as well as number ad nature of dimensions. The empirical results show that the proposed framework is indeed sufficiently general and reusable [34].

Cyber security in the context of big data is known to be a critical problem and presents a great challenge to the research community. Sabar et al. have proposed a novel, domain-independent hyper-heuristic framework for the formulated "Support Vector Machine" (SVM) configuration process as a bi-objective optimization problem in which accuracy and model complexity are considered as two conflicting objectives. The effectiveness of the proposed framework has been evaluated on Microsoft malware big data classification and anomaly intrusion detection [35].

The next work analyzes the ability of popular selection operators used in a hyper-heuristic framework to continuously select the most appropriate optimization method over time. Van der Stockt et al. have presented the considerations and criteria to select a diverse mix of heuristics specific to dynamic optimization problems and non-dynamic optimization problems to enable the heterogeneous meta-hyper-heuristic to effectively solve dynamic optimization problems [36].

A review article identifies the characteristics necessary for the development of frameworks for optimization using metaheuristics and, from these characteristics, identifies existing gaps, especially those related to the hybridization of metaheuristics. Silva et al., in this article, have also showed that the concepts of multi-agent systems in the design of frameworks for optimization using metaheuristics facilitates and flexibilizes the development of hybrid metaheuristics and allows simultaneous exploration of different regions of the search space [37].

A Modified Choice Function (MCF), a hyper-heuristic method, is applied such that it can regulate the selection of the neighborhood search heuristics adopted by the employed and onlooker bees automatically. The proposed MCF-ABC (Artificial Bee Colony) model is a bee algorithm with multiple neighborhood search heuristics. While the employed bees and onlooker bees perform neighborhood search to exploit the promising areas of the search space, the scout bees focus on exploration of a new region in the search space. The proposed model solves the 64 Traveling Salesman Problem instances available in TSPLIB, on average, to 0.055% from the known optimum within approximately 2.7 minutes [38].

Ahmed et al. have evaluated the performance of a set of selection hyper-heuristics on the route design problem of bus networks, with the goal of minimizing the passengers' travel time, and the operator's costs. Their analysis shows the success of the sequence-based selection method combined with great deluge acceptance method, outperforming other selection hyper-heuristics in both passenger and operator objectives [39].

A hyperheuristic framework, namely hyperSPAM, composed of three search algorithms for continuous optimization problems has been proposed. The main focus is to select the search algorithms correctly such that a simple random coordination can lead to satisfactory results. Four coordination strategies, in the fashion of hyperheuristics, have been used to coordinate the second and the third single-solution search algorithms. One of them is a simple randomized criterion while the other three are based on a success based reward mechanism [40].

Lin has proposed an effective backtracking search based hyper-heuristic (BS-HH) approach to address the Flexible job-shop scheduling problem (FJSP) with fuzzy processing time (FJSPF). A back-tracking search algorithm is introduced as the high-level strategy to manage the low-level heuristics incorporated into the BS-HH to operate on the solution domain directly. Additionally, a novel hybrid solution decoding scheme is proposed to find an optimal solution more efficiently. The author has emphasized by saying that the FJSPF which extends FJSP by allowing processing time or due date to be fuzzy variable is more close to the real-world situation [41].

A two-layered decision-making system is proposed where the first step introduces task allocation and sequencing into the system's energy management procedures for the purpose of enabling long-term autonomy of a heterogeneous of marine robots, and the next step includes constructing a validation and evaluation system for solutions and methods which will enable objective grading during the process of training a hyper-heuristic top decision-making layer [42].

Another study evaluates Multi-Objective Agent-Based Hyper-Heuristic in real-world applications, by searching solutions for four multiobjective engineering optimization problems. For this purpose, an additional multi-objective evolutionary algorithm and new quality indicators better adapted to real-world problems are used [43].

Next, the resulting sequences of low level heuristic selections and objective function values minimized through the use of a selection hyper-heuristic are used to generate a database of heuristic selections. The sequences in the database are broken down into subsequences and the mathematical concept of a logarithmic return is used to discriminate between "effective" subsequences, which tend to decrease the objective value, and "disruptive" subsequences, which tend to increase the objective value. These subsequences are then employed in a *sequenced based hyper-heuristic* and evaluated on an unseen set of benchmark problems [44].

Motivation in another work is to generate effective dynamic scheduling policies (SPs) through off-line learning and to implement the evolved SPs online for fast application. Three types of hyper-heuristic methods are proposed for coevolution of the machine assignment rules and job sequencing rules to solve the multi-objective dynamic flexible job shop scheduling problem. The results reveal that the evolved SPs can discover more useful heuristics and behave more competitive than the man-made SPs in more complex scheduling scenarios without increasing the online solution time. It also demonstrates that the evolved SPs can obtain trade offs among different objectives and have a strong generalization performance to be reused in new unobserved scheduling scenarios, which make the evolved SPs more robust when they are employed in a stochastic and dynamic scheduling environment [45].

A case study focusing on multi-objective flexible job shop scheduling problem (MO-FJSP) in an aero-engine blade manufacturing plant has been proposed. Three multi-agent-based hyper-heuristic integrated with the prior knowledge of the shop floor are proposed to evolve SPs for the online scheduling problem. This situation poses a major challenge to the current scheduling system because dynamic changes in the shop-floor require real-time responses thanks to the widely used numerous sensors, automatic robots and enhanced systems in it. Since it is not necessary to get an optimal solution in real-time scheduling, the use of heuristics to produce a satisfactory solution for solving the production scheduling problem in an acceptable time frame is a viable option. With the change of orders, routes and other elements in the shop-floor, the previously established rules may not be able to adapt to new scheduling scenarios. Hence, the implementation of hyper-heuristics to further enhance the heuristics made by experts is necessitated. The results show that the bottleneck agent model is more favorable than the other two agent models and succeeds to make a good trade-off between the solution quality and the generalization performance among the three agent models [46].

Oyebolu et al. have presented a discrete-event simulation of continuous bioprocesses in a scheduling environment. More specifically, characteristics specific to bioprocessing and biopharmaceutical manufacturing are addressed using a simulation optimization approach. For the optimization algorithm, the authors use an evolutionary algorithm to search for optimal production control policies. As they search the space of possible heuristics or rules as opposed to possible solutions, it constitutes a hyper-heuristic approach. Dynamic SPs are investigated to make operational decisions in a multi-product manufacturing facility and react to process failure events and uncertain demand. In particular, tuning of process run times leads to improved performance as this enables better lot-sizing decisions which may allow hedging against process failure by utilizing a shorter run time [47].

Leng et al. have investigated the optimization of a variant of the location-routing problem (LRP), namely the regional low-carbon LRP (RLCLRP), considering simultaneous pickup and delivery, hard time windows, and a heterogeneous fleet. In order to solve this problem, the authors construct a biobjective model for the RLCLRP with minium total cost consisting depot, vehicle rental, fuel consumption, carbon emission costs, and vehicle waiting time. They also further propose a novel hyper-heuristic method to tackle the biobjective model. The proposed method applies a quantum-based approach as a high-level selection strategy and the great deluge, late acceptance, and environmental selection as the acceptance criteria [48].

Finally, Lissovoi have aimed to extend the understanding of the behavior and performance of hyper-heuristcis to multimodal optimization problems. In order to evaluate their capability at escaping local optima, they consider elitist, which only accepts moves that improve the current solution, and the non-elitist, which accepts any new solution independent of its quality, selection operators that have been used in hyper-heuristics in the literature [49].

3. Heuristically generated angular metric approach (HAMA)

Three heuristic properties are important. The first one is a property of dominance as a distinctive factor. The objective here is to increase the discriminativeness, because the efficiency of the heuristic entirely depends on it. The second one is the property of consistency which is based on the triangle inequality. The sum of the two sides of the three that make up a triangle must always be greater than or equal to the third side; their difference must also be less than or equal to the third one. For example, suppose we are on the node 20 at any puzzle problem. The functional value of the sum of the twentieth node and its successor must be larger than the one arising from the successor node so that the heuristic function should not repeat in itself. The third property is admissibility. In this property, any distance determined by a heuristic function is always expected to result in smaller than the actual distance. In other words, heuristic will never yield an overestimation, thereby making it an admissible one.

In light of these three properties above, it is evident that the metrics turn into a straight line when it starts from a semi-circle. The fact that it is a semi-circle is consistent with explaining the principle brought about by triangle inequality. It is possible to further state that a heuristic acquisition is proportional to the arc length of the circle-segment in geometrical representation of HAMA [50].

Although the triangle has different line-segments (i.e., sides) such as r_1 and r_2 in the triangle (see **Figure 1**), the heuristic acquisition can be explained by setting up a proportional relation with the arc length seen by the angle of α of an isosceles triangle which is formed by selecting the smaller of r_1 and r_2 as the equal sides of the triangle and the radius of the circle as shown in **Figure 1**. Moreover, it should be noted the appearance of the angle of α as a choice which is not dependent on the lengths of the sides given by r_1 and r_2.

The arc length of the circle-segment shown in **Figure 1** is equal to $2\pi r\alpha/360^\circ$ where α resides in the range of $0^\circ \leq \alpha \leq 180^\circ$.

We will focus specifically on the Manhattan, the Euclidean, and the Chebyshev as a common distance metrics/measures. These distance metrics have advantages and pitfalls depending on when and how they will be used. One may emerge as a better alternative to the other. For example, k-NN, a technique often used for supervised learning, often uses the Euclidean metric. But it would not work if our data is high dimensional or consists of geospatial information.

The Euclidean distance metric: We can easily calculate the distance from the Cartesian coordinates of the points according to the Pythagorean theorem:

$$D_E(x,y) = \sqrt{\sum_{i=1}^{n} (x_i - y_i)^2} \tag{1}$$

Two major disadvantages of the Euclidean distance metric are the lack of scale-invariance property and the dimensionality increase of the data used. Although it is a common distance metric, the Euclidean distance metric becomes the less useful under these situations. In particular, due to the *curse of dimensionality*, high dimensional space does not behave intuitively as desired in 2- or 3-dimensional space. The more the dimension increases, the closer will be the average distance and the maximum distance between randomly placed points. Therefore, the Euclidean distance metric works well in the case of low-dimensional data where it is important to measure the magnitude of the vectors.

The Manhattan distance metric: It refers to the distance between two vectors that can only move at right angles to each other. Diagonal movement is not taken into account when calculating the distance:

$$D_M(x,y) = \sum_{i=1}^{k} |x_i - y_i| \tag{2}$$

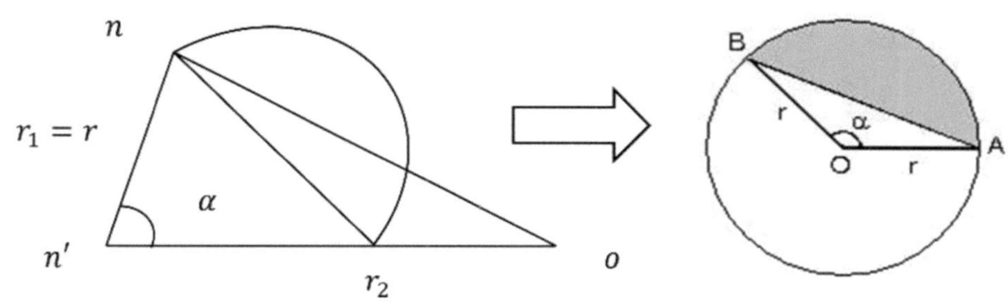

Figure 1.
Side-selection to determine the circle-segment proportional to heuristic acquisition (shaded area).

It is a less intuitive metric than the Euclidean metric, especially when used with high-dimensional data. If the dataset to be used has discrete and/or binary attributes, the Manhattan stands out as a metric that works quite well since it takes into account the realistic paths that would be taken within those attribute values. For example, while the Euclidean metric could create a straight line between two vectors, it is highly unlikely that in reality this is actually possible!

The Chebyshev distance metric: The Chebyshev distance determines the largest difference between two vectors along any coordinate dimension. With a more rigorous definition, it represents the maximum amplitude difference of the two vectors in terms of coordinates. In other words, it is simply the maximum distance through one axis:

$$D_C(x,y) = \max_i \left(|x_i - y_i|\right). \tag{3}$$

It can also be a useful metric in games that allow unrestricted 8-way movement. In warehouse logistics, it is preferred because the time an overhead crane takes to move an object is similar to the Chebshev distance. Except in such very special cases, it is unlikely to be used as an all-purpose measure of distance, such as the Euclidean or the Cosine similarity.

All the angles in the range of $0^\circ \le \alpha \le 180^\circ$ are within the limits of the *heuristic metric approach*, *HAMA*. In this range, an unlimited number of metrics can be achieved including the Manhattan, the Euclidean, and the Chebyshev distance metrics covered [50]. For example, if α is selected 120 degrees, then the heuristic distance metric between the two vectors under the assumption of equal segments, that is, equal lengths of adjacent nodes n' and n and n' and 0, represented as $\Delta x = \Delta y$ (that is, $|x_1 - x_2| = |y_1 - y_2|$, respectively, since our approach accepts equal sides as circle radius) can be calculated by using the Cosine theorem as follows:

$$\sqrt{(\Delta x)^2 + (\Delta y)^2 - 2\Delta x \Delta y cos 120^\circ} = \sqrt{3}\Delta x. \tag{4}$$

If α is selected 60 degrees, then the new heuristic distance metric can be calculated as,

$$\sqrt{(\Delta x)^2 + (\Delta y)^2 - 2\Delta x \Delta y cos 60^\circ} = \Delta x. \tag{5}$$

On the other hand, the Manhattan distance metric will calculate the result as follows:

$$|x_1 - x_2| + |y_1 - y_2| = 2\Delta x. \tag{6}$$

The same result can also be found with the Cosine theorem by taking the angle 180 degrees:

$$\sqrt{(\Delta x)^2 + (\Delta y)^2 - 2\Delta x \Delta y cos 180^\circ} = 2\Delta x. \tag{7}$$

Similarly, the Cosine theorem with an angle of 90 degrees and the Euclidean metric will give the same result of $\sqrt{2}\Delta x$.

Another equal result is obtained as Δx between Chebshev and the Cosine theorem with an angle of 60 degrees. This result is misleading. In order for HAMA to produce the correct results is that the lengths between each pair of nodes, that is, the lengths of the segments, have to be given differently (i.e., $r_1 \neq r_2$) from each

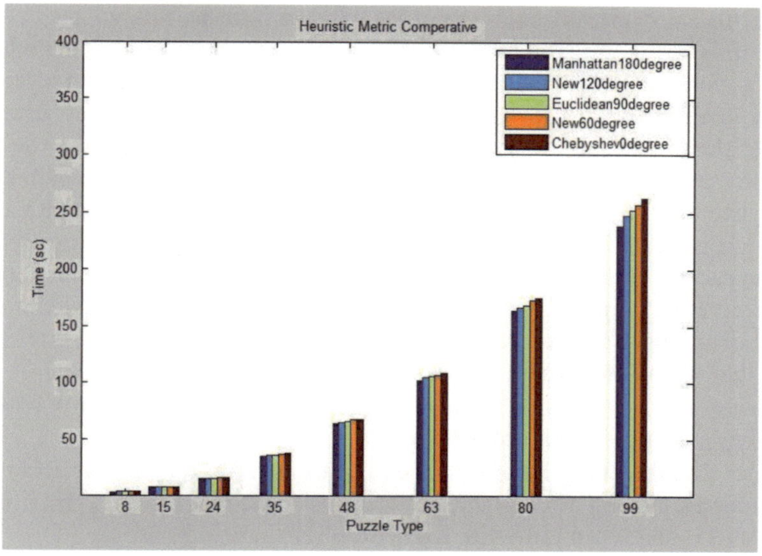

Figure 2.
A comparison of different heuristic metrics.

other in the original problem. In fact, in general, the Cosine theorem is used to solve for missing side or angle in non-right triangle and a non-matching pair/side. As we defined earlier, the Chebyshev distance metric is obtained from the maximum amplitude difference among elements of the two vectors for a given dimension. However, this amplitude-based distance is sensitive to spikes or abnormal peaks in data. As we already know, the angular distance metric belongs to the family of cosine distances derived from the Cosine similarity metric. As long as the angle between the vectors under consideration is maintained, the major advantage of the angular distance metric is its low sensitivity to any changes in vector norms, thus providing the desired distance that is not dependent on the amplitude [51].

The computer used in the simulations features an Intel 4-core i5-3230 processor with a 2.6 GHz CPU clock rate and 8 GB RAM. Run times toward a solution for each one of the metrics in 8-puzzle, 15-puzzle, 25-puzzle, 35-puzzle, 48-puzzle, 63-puzzle, 80-puzzle, and 99-puzzle problems have been compared and the comparisons have been presented in **Figure 2**.

4. An application of heuristically generated distance-metric to a chase problem

4.1 Problem statement

In Cat chasing mouse (CCM) problem, the actions and reactions of the pursuer and prey are designed to be as realistic as possible to the real-world. The prey (i.e., mouse) is smart enough how to avoid the pursuer's (i.e., cat's) maneuvers. The model allows multiple cats to chase a single mouse. Cats know how to take the appropriate angles to block off the mouse [52].

For example, the mouse's decision-making procedure works like this: the strategy is developed in such a way that the mouse's future route is to distance itself from the nearest cat where it sees the greatest threat. After determining which cat is closest, the mouse runs in the direction that the cat is running. If the cat is away from the mouse, it is not always necessary to do so away from the cat.

Cat's artificial intelligence, on the other hand, is built on fuzzy logic control (FLC), mainly to increase the cat's ability to catch the mouse [53]. The model checks if the cat is close enough to the mouse. If close, the cat uses the FLC to set its course. If it is far, it sets its course for a point in front of the mouse using the current position of the mouse and the angle it is facing. If the cat is away from the mouse but the mouse is partially moving towards the cat, then the route is again determined by the FLC. This model also allows the cats close to the mouse to move in a group, while the cats far away from the mouse block off the mouse in case of an attempt to escape. Using the random cat locations has also been a good choice for the model to better represent a real-world situation. Finally, the Max-Min rule is preferred over Kosko's Max-Product rule as the inference method in this model, claiming that it leads to a more successful FLC [52].

At this point, x_{cat} and y_{cat} represent the coordinates of cat's position whereas x_{mouse} and y_{mouse} represent the coordinates of mouse's position.

Let us define the mouse's speed traveling eastwards as a fixed variable v, and the cat's speed traveling at a pursuit direction as a fixed variable w, and it is always $w > v$. From **Figure 3**, we can write from AOB right-angled triangle,

$$\tan\left(azi(t) + ang(t)\right) = \frac{x_{mouse} - x_{cat}}{y_{mouse} - y_{cat}}. \tag{8}$$

Applying the inverse tangent (functional) operator to each side of Eq. (8) yields,

$$ang(t) = atan\left(\frac{x_{mouse} - x_{cat}}{y_{mouse} - y_{cat}}\right) - azi(t), \tag{9}$$

where $ang(t)$ is a dependent dynamic variable.

4.2 Fuzzy logic control (FLC) and design procedures for cat-mouse problem

4.2.1 Mathematical model

First of all, we will set up a mathematical model of the "plant" and determine the state variables and control input variables from this model as follows:

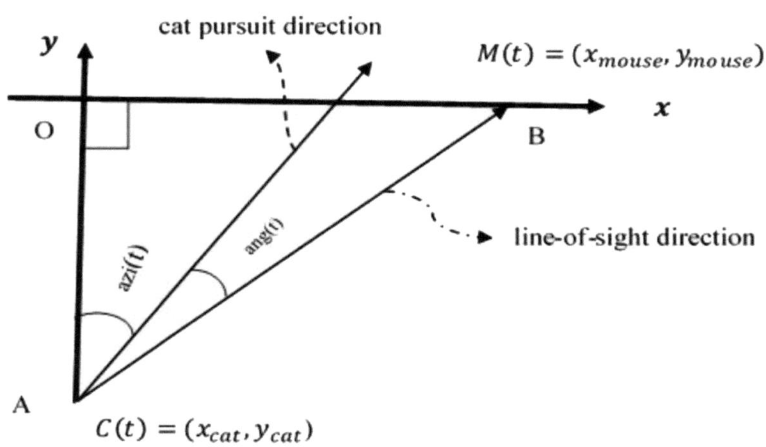

Figure 3.
Geometric representation of the problem.

$$
\begin{aligned}
x_{mouse}(t+1) &= x_{mouse}(t+1) + v.1; \\
y_{mouse}(t+1) &= y_{mouse}(t); \\
x_{cat}(t+1) &= x_{cat}(t) + \text{w.}\sin(azi(t+1)); \\
y_{cat}(t+1) &= y_{cat}(t) + \text{w.}\cos(azi(t+1)); \\
azi(t+1) &= azi(t) + dz(t).1;
\end{aligned}
\tag{10}
$$

where state variables are mouse's and cat's positions, that is, $M(t) = (x_{mouse}(t), y_{mouse}(t))$ and $C(t) = (x_{cat}(t), y_{cat}(t))$, respectively, and azimuth angle $azi(t)$. Control input variable is $dz(t)$. We will keep these state variables as crisp variables. On the other hand, the control input $dz(t)$ and dependent dynamic variable $ang(t)$ which is derived from the state variables will be characterized as fuzzy variables that will be fuzzified to apply the fuzzy logic.

Next, we will specify the control objective. Our control objective is to minimize $ang(t+1)$ by applying appropriate control input $dz(t)$ to update $azi(t+1)$ as $ang(t)$ is given. Then, we can express our control law as part of the control effort as follows:

$$dz(t+1) = K.ang(t), \tag{11}$$

which is simply 'proportional control' P. In this control law, how to choose the most appropriate value of the proportional constant K is the main question we seek the answer to. Since this system is observed as nonlinear by nature, conventional PID design technique could be difficult to apply. Therefore, extensive simulation and trial-and-error would be required.

4.2.2 Fuzzy logic control (FLC) architecture

Fuzzy Logic Control (*a.k.a.* Fuzzy Linguistic Control) is a knowledge based control strategy that can be used when either a sufficiently accurate and yet not unreasonably complex model of the plant is unavailable, or when a (single) precise measure of performance is not meaningful or practical.

FLC design is based on empirically acquired knowledge regarding the operation of the process. This knowledge, transformed or changed into linguistic, or rule-based form, is the core of the FLC system. FLC architecture differs from a conventional controller in that the selected control algorithm and mathematical model blocks in the conventional controller will be completely replaced by Fuzzifier (Encoder), Defuzzifier (Decoder), Knowledge, and Inference Engine as shown in **Figure 4**.

In this architecture, the dynamic filter computes all the system dynamics: x (state variables) consists of selected elements of $e = r - y$, de/dt, and $\int e d\tau$. The rule base (knowledge base) provides nonlinear transformation without any built-in dynamics.

4.2.3 Fuzzification of the input variable 'ang' to fuzzy logic controller

Fuzzification is the process of decomposing a range of each fuzzy variable into one or more fuzzy sets via membership functions. Simply, it converts crisp sets to a fuzzy set. To achieve the fuzzy decomposition firstly, the range of each fuzzy variable will be specified, and then a set linguistic variables will be devised. Linguistic variables are variables whose values are words in natural language.

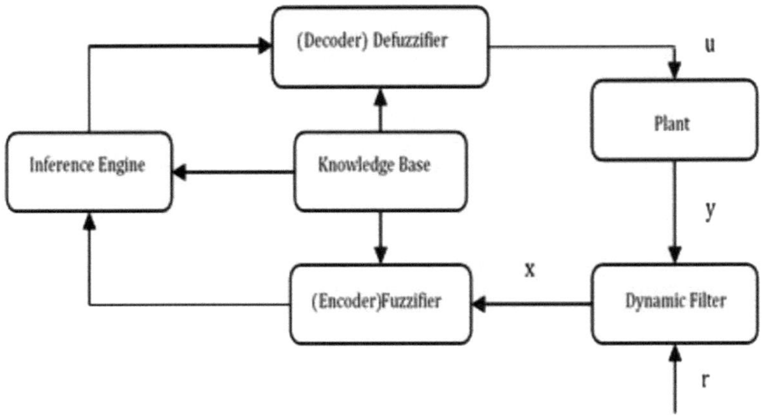

Figure 4.
FLC architecture.

Fuzzification is no different from finding an estimate of an input value. That is, it returns an *activation vector* as a fuzzy function output, representing linearly interpolated values of a given input vector '$ang(t)$'. In the meantime, briefly the *activation vector* is a row vector of the same size as the number of columns of a fuzzy set matrix M. If '$ang(t)$' falls outside the range of the support vector, an error is reported. If '$ang(t)$' falls between two elements of the support vector, a linear interpolation is performed as follows:

Let us assume that we have two known points x_1, y_1 and x_2, y_2. Our objective is to estimate y value for some x value that is between x_1 and x_2. We call this y value an "interpolated" value. There exists two simple methods for choosing y. The first one is to see whether x is closer to x_1 or to x_2. If x is closer to x_1, then we use y_1 as the estimate, otherwise we use y_2. This is called 'nearest neighbor' interpolation. The second one is to draw a straight line between x_1, y_1 and x_2, y_2. We look to see the y value on the line for our chosen x. This is called "linear interpolation."

Straight-line equation between x_1, y_1 and x_2, y_2 in **Figure 5** is given as:

$$y = y_1 + (x - x_1)\frac{y_2 - y_1}{x_2 - x_1} \tag{12}$$

where y is the estimate of the chosen value x, and represents the output of the 'fuzzify' function 'a'.

Firstly, between $ang(t)$ and 'support for $ang(t)$', $sang$, which we will define later, the following relationship exists: $ang(t) = sang(ne)$. At the output, the following relationship will be observed: $a = M(:, ne)^T$, where 'ne' is the initials of 'number of elements'.

Next, let '$ang(t)$' be our x. In this case, if we assume that $y_1 = M(:, ne)^T$, then, $y_2 = M(:, ne+1)^T$. In response to these two, we can write the followings as a result, respectively: $x_1 = sang(ne)$ and $x_2 = sang(ne+1)$. If we place all of these into the straight-line equation in Eq. (12), then we can write:

$$a = M(:, ne)^T + (ang(t) - sang(ne))\frac{M(:, ne+1)^T - M(:, ne)^T}{sang(ne+1) - sang(ne)}. \tag{13}$$

Here, if a suitable definition is made, such as $alpha \triangleq \frac{(ang(t) - sang(ne))}{sang(ne+1) - sang(ne)}$, which simplifies the above expression, then we get:

$$a = M(:,ne)^T + alpha\left(M(:,ne+1)^T - M(:,ne)^T\right). \quad (14)$$

By rearranging above, we can finally write the following:

$$a = alpha * M(:,ne+1)^T + (1 - alpha) * M(:,ne)^T. \quad (15)$$

This appears as the most suitable form of equation for coding the chosen programming language.

Fuzzification uses a discrete support. The universe-of-discourse (UoD) (i.e., the range of all possible values applicable to the chosen variable) of support are sampled at uniform (or non-uniform) intervals. For our CCM problem, both '$ang(t)$' and '$dz(t)$' will share the same set of linguistic variables: LN, SN, ZO, SP, LP. The dynamic ranges of the supports (i.e., UoD) of these two fuzzy variables however, are different: $ang(t) = -180^0$ *to* 180^0 and $dz(t) = -30^0$ *to* 30^0. These choices may be due to physical constraints and other prior knowledge. In this problem, the range difference is determined by the proportional control (law) constant K which were set to 1/6 [53]. Therefore, $dz = K.ang$, and $sdz = K.sang$. It is important that the supports of adjacent linguistic variables overlap so that more than one fuzzy rules may be fired as shown in **Figure 6**.

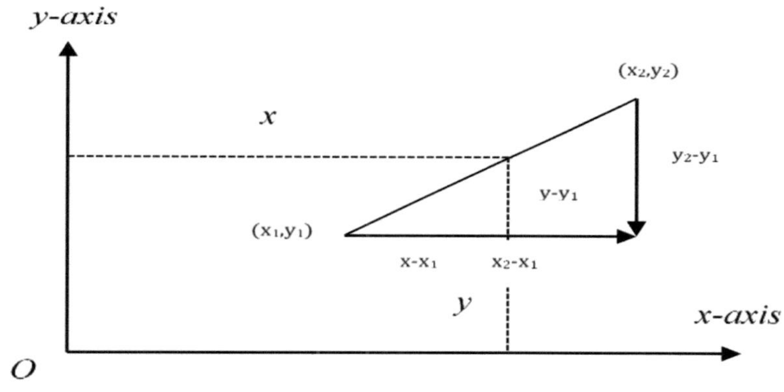

Figure 5.
Linear interpolation by drawing a straight-line between two points.

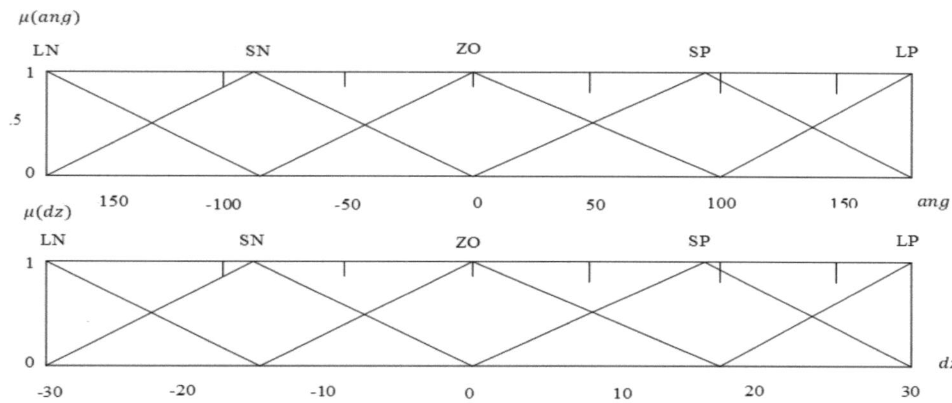

Figure 6.
Fuzzy quantization of the state variables into a set of linguistic variables. Output: fuzzy inputs. Inputs: Membership functions vs. crisp inputs.

Fuzzification is essentially a fuzzy quantization of the state variables. In this respect, the state variables '*ang*' and '*dz*' may be quantified into a set of linguistic variables, with two parameters, polarity and size, as follows: NL-Negative, Large; NS-Negative, Small; ZO-Zero; PS-Positive, Small; PL-Positive, Large. Fuzzification process converts a crisp sensor reading (value of state variable) $x = x_0$ into the grade values of each of these linguistic variables. In particular, we have,

$$[\mu_{NL}(x_0), \mu_{NS}(x_0), \mu_{ZO}(x_0), \mu_{PS}(x_0), \mu_{PL}(x_0)].$$

Fuzzy sets other than LN- and LP- consist of arrays of elements on the identical sides of an isosceles triangle, moving in the same direction at uniform intervals (i.e., quantized values of linguistic variables) and ultimately forming the row of the M fuzzy matrix. On the other hand, LN- and LP-fuzzy sets are the arrays whose elements consist of the points advancing at uniform intervals on only one of the identical sides of a triangle as follows:

$$M = \begin{bmatrix} 1 & 0.67 & 0.33 & 0 & 0 & 0 & 0 & 0 & 0 & 0 & 0 & 0 & 0 \\ 0 & 0.33 & 0.67 & 1 & 0.67 & 0.33 & 0 & 0 & 0 & 0 & 0 & 0 & 0 \\ 0 & 0 & 0 & 0 & 0.33 & 0.67 & 1 & 0.67 & 0.33 & 0 & 0 & 0 & 0 \\ 0 & 0 & 0 & 0 & 0 & 0 & 0 & 0.33 & 0.67 & 1 & 0.67 & 0.33 & 0 \\ 0 & 0 & 0 & 0 & 0 & 0 & 0 & 0 & 0 & 0 & 0.33 & 0.67 & 1 \end{bmatrix}$$

Furthermore, from **Figure 6**, we can see that the ZO and SP fuzzy sets are the shifted versions of the SN fuzzy set, which is formed by moving from the left zero to the right zero of the triangle in **Figure 7**. After drawing attention to these important points, we can now construct the M fuzzy matrix as follows:where the first row is the LN fuzzy set, the second row is the SN fuzzy set, and similarly the third row is the ZO fuzzy set, the fourth row is the SP fuzzy set, and finally the fifth row is the LP fuzzy set.

4.2.4 Representing fuzzy logic control rules

Rules can be represented conveniently as a matrix if there are two input fuzzy variables. To represent rules, a matrix is defined where each row is a rule. Each row contains $d(1) + d(2)$ elements. Here, $d(1)$ represents the number of fuzzy sets defined on the input variable and $d(2)$ represents the number of fuzzy sets defined on the output variable. The first $d(1)$ elements specify which fuzzy set is used as the antecedent part. The next $d(2)$ elements specify which fuzzy sets in the output control variable are used. A simple rule base for the CCM problem that has only one input fuzzy variable '$ang(t)$' and one control (output) variable '$dz(t+1)$'.

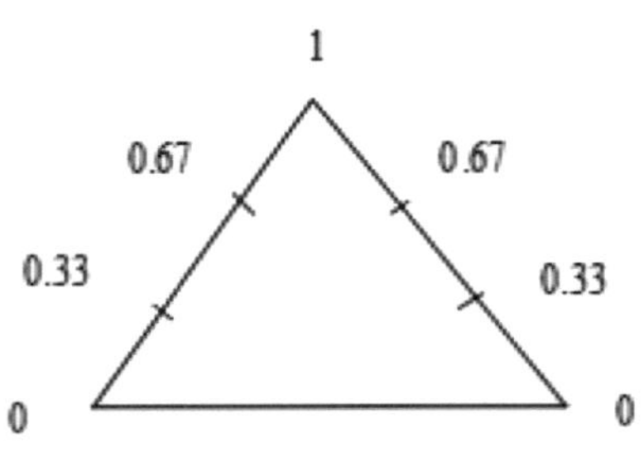

Figure 7.
Fuzzy sets arising as arrays of elements at uniform intervals.

Rule matrix is defined as follows:

$$
rule = \begin{bmatrix}
\overbrace{\begin{matrix} LN & SN & ZO & SP & LP & \&? & LN & SN & ZO & SP & LP & weight \end{matrix}} \\
\begin{matrix}
1 & 0 & 0 & 0 & 0 & 1 & 1 & 0 & 0 & 0 & 0 & 1 \\
0 & 1 & 0 & 0 & 0 & 1 & 0 & 1 & 0 & 0 & 0 & 1 \\
0 & 0 & 1 & 0 & 0 & 1 & 0 & 0 & 1 & 0 & 0 & 1 \\
0 & 0 & 1 & 0 & 0 & 1 & 0 & 0 & 1 & 0 & 0 & 1 \\
0 & 0 & 0 & 0 & 1 & 1 & 0 & 0 & 0 & 0 & 1 & 1
\end{matrix}
\end{bmatrix}.
$$

In the rule matrix, there are five rules, one from each row, respectively:

$$\text{If } ang(t) \text{ is LN/SN/ZO/SP/LP, then } dz(t+1) \text{ is LN/SN/ZO/SP/LP.}$$

$\&? = 1$ means that if there is only one input fuzzy variable (i.e., this case) or the second fuzzy variable is to be ignored for that rule. The last column indicates the relative weighting of that rule.

4.2.5 Inference engine

A fuzzy inference engine is a mechanism to calculate an output from given inputs using fuzzy logic. When input variables are fuzzified, each rule in the rule base will try to determine its degree of activation using '*min-max*' or '*correlation-max*' method. For rules which have a non-zero activation value, the output fuzzy variables will be combined (fuzzy union) yielding a resultant fuzzy set.

Now, let us show how to create a rule matrix where each row corresponds to a rule, then its rows and columns can be specified by '*nrule*', and $(d(1) + d(2) + d(3) + 1)$, respectively.

Each row consists of five parts. The first part with $1xd(1)$ dimensional input variable '$input1$' and the second part with $1xd(2)$ dimensional input variable '$input2$' are allocated for antecedent variables. The third part, which is a single column, will be equal to '1' if '$input2$' is to be ignored for that rule, and '0' otherwise. The fourth part with $1xd(3)$ dimensional output variable, *output*, is fuzzy representation of consequent variable. Here, $d = [d(1)\, d(2)\, d(3)]$ is a 1×3 dimensional vector that specifies the number of fuzzy sets (adjectives or linguistic variables) defined on each UoD. In this d vector, $d(2)$ gives the number of fuzzy sets of the second input variable, if there are 6 input arguments for inference function, otherwise (i.e., the case of 5 input arguments only for the function), $d(2)$ gives the number of fuzzy sets of the output, or else $d(3)$ is the number of fuzzy sets defined on the output variable.

- First, we consider the first $d(1)$ columns of the matrix giving the first input set.

- Then, an $nrulex1$ column vector, of which all elements are '1', premultiplied by a $1xd(1)$ row vector 'a_1', is multiplied by the corresponding elements of the first $d(1)$ columns of the rule matrix to eventually produce a matrix of size $nrulexd(1)$. Here, 'a_1' represents the activation of fuzzy (antecedent) variables for input-1 and has the same dimension as $d(1)$.

- Next, we apply the 'max' operation to the $nrulexd(1)$ matrix after it is transposed, find the maximum values in each '*nrule*' column, and place these values in a $1xnrule$ row vector.

- Finally, the *nrule* dimensional row vector is converted back to an *nrule* dimensional column vector with a transpose operation and is represented by A_1.

Case 1. Having only input-1 or $d(1)$:

- *This nrulex*1 *column vector A_1 is multiplied by the corresponding elements of the same size column vector called 'weight' which is positioned in the last column of the rule matrix and represents the weighting of each rule.*

- *Each element of this nrule-by-1 product is placed on the main diagonal of an nrule dimensional square matrix which is also a diagonal matrix.*

- *Later, this diagonal matrix is multiplied by the output fuzzy variable set.*

Case 2. Having the second input, input-2 or $d(2)$, as well:

- *First, we consider the next $d(2)$ columns of the matrix giving the second input set.*

- *Then, an nrule-by-1 column vector, of which all elements are '1', premultiplied by a* 1x$d(2)$ *row vector 'a_2', is multiplied by the corresponding elements of the next $d(2)$ columns to eventually produce a matrix of size nrulexd(2). Here, 'a_2' represents the activation of fuzzy (antecedent) variables for input-2 and has the same dimension as $d(2)$.*

- *Next, we apply the 'max' operation to lastly obtained nrule-by-$d(2)$ matrix after it is transposed, and as a result, we get an nrulex*1 *column vector, A_2.*

- *A 'min' operator is then applied on a* 2*xnrule transposed augmented nrulex*2 *matrix of A_1 and A_2. The result is a* 1*xnrule row vector consisting of the minimum elements in each column of the transposed augmented matrix of $[A_1 : A_2]^T$.*

- *Afterwards, this* 1*xnrule row vector converted to nrulex*1 *column vector via a transpose operator is multiplied by the corresponding elements of the column vector 'weight' which forms the last column of the rule matrix. The resulting product is also an nrulex*1 *column vector.*

- *In addition, each element of this column vector is placed on the main diagonal of an nrule dimensional square matrix which is also a diagonal matrix.*

- *Besides, this diagonal matrix is multiplied by a submatrix consisting of columns that make up the output variable of the rule matrix. Let us call this nrule-by-dout matrix 'Tmp' which gives the activation of each rule. Here, 'dout' represents the number of the fuzzy sets defined on the output variable.*

- *Now, with the 'max' operator to be applied on this nrulexdout matrix 'Tmp', we will obtain a row vector whose elements consist of the maximum values in each column of the aforementioned matrix. This dout dimensional row vector called 'act' will represent the activation of each output fuzzy set.*

- *B is an output fuzzy variable matrix whose size is defined by dout-by-nofz, where dout is equal to $d(2)$ if there is a single input, and $d(3)$ in case of two inputs, and nofz is equal to the length of a row vector giving the coordinates of UoD, or it is equal to the number of output fuzzy sets which is a subset of UoD. In other words, each row* *in B is a fuzzy set defined on that output variable and each column is the element of the discrete support it corresponds to.*

- *If Kosko's max product rule is applied, then the output is calculated with the following formula where "*" implies that the product rule is already applied:*

$$out = max\left(diag(act) * B\right). \tag{16}$$

- *If the max-min rule will be applied instead of the product rule, the calculation will then be as follows:*

    - *Firstly, the 'act' vector converted into a doutx1 column vector is multiplied by a 1xnofz row vector composed of 1 s equal to the number of the output fuzzy sets.*
    - *The objective here is to generate a sparse matrix of the same size as the B matrix to be compared via the 'min' operator.*
    - *According to the max-min rule, firstly, by applying the 'min' operator between this generated sparse matrix and the B matrix, an array of the same size as the sparse matrix and the B matrix is created, in which the elements of both matrices in the same position are compared and the smaller one is placed.*
    - *Finally, by applying the 'max' operator to the resulting matrix, a row vector containing the maximum element from each column will be returned as the output 'out'.*
    - *The equation of the max-min rule applied step by step above is as follows:*

$$out = max\left(min\left(act^T * ones(1, nofz), B\right)\right). \tag{17}$$

4.2.6 Defuzzification

Fuzziness helps us to evaluate the rules, but the final output of a fuzzy system has to be a crisp number. Defuzzification is the process of combining the successful fuzzy output sets produced by the inference mechanism. The input for the defuzzification process is the aggregate output (i.e., unified outputs of all rules) fuzzy set '*out*'.

Most popular defuzzification method is the *centroid technique*. It involves calculating the point where a vertical line would slice the aggregate set into two equal masses. It calculates the centroid of output of the fuzzy set '*out*' defined on the discrete support, '*support*'. Here '*out*' and '*support*' should have the same size. The *centroid method* equation is:

$$y = \frac{\int \mu_B(z).z\, dz}{\int \mu_B(z) dz} \tag{18}$$

Centroid defuzzification method finds a point representing the center of gravity (or area) of the fuzzy set B on the interval ab. In Eq. (18), z represents a crisp sensor reading or value of the state variable, and $\mu_B(z)$ is the fuzzy quantity that is the graded value of the particular linguistic variable upon the process of conversion (i.e., fuzzification process) from the crisp or precise quantity. Simply, we can adapt Eq. (18) for coding convenience with our values '*out*' and '*support*' as follows:

$$y = sum(out. * support)/sum(out) \tag{19}$$

5. Numerical experiments and simulations

In this work, we have devised a heuristic distance metric approach that employs the Cosine theorem, which uses the angle between the line-of-sight direction from cat to mouse and the cat's current pursuit direction only at non-right angles, rather than the distance metric which calculates the distance of current cat to mouse based on the Euclidean distance [52]. As here, if the metric that calculates the current distance from cat to mouse is not always taken in Euclidean, but built around our heuristic metric approach that allows an unlimited number of metrics to be achieved, whereby the distinction between the computation times of each metric can be clearly seen in **Figure 2** as the puzzle size increases, mouse will gain more time in perceiving the incoming danger, thus increasing the percentage of evading it, and will escape.

We have also compared our heuristic approach with the metric distance between two adjacent nodes, d_m, that do not take into account the angle between the nodes, defined as [54],

$$d_m = \frac{l_i + l_j}{2}, \{l_i, l_j\} \in R^+, \tag{20}$$

where l_i, l_j represent the metric lengths of the segments i and j, respectively. Here, the metric distance defines geodesics (i.e., the shortest paths) as those paths with minimal sum of metric length.

Next, we chose randomly generated cat locations as they better reflect the real-world in each run, and compared the results by running a total of 100 trials for each of the selected cases that accepted first 4 cats and then 8 cats and finally 12 cats against a single mouse. Thus, 'caught and escape percentages vs. number of cats' findings for three metric distances (i.e., the metric calculated only based on the EUCLIDEAN distance, the metric calculated over the GEODESICS definition, and the metric devised using the angle-dependent Cosine theorem as a result of our angular metric approach, that is HAMA) have been searched for and the results evaluated comparatively. **Table 1** below confirms that our approach, which we call HAMA, performs best in terms of evasion and escape performance that we consider:

As can be seen from **Figure 8**, the best caught and escape performance is exhibited by HAMA. That is, when compared to the other two metrics, the caught performance is minimum and the escape performance is maximum. Another note-worthy finding is that as the number of predators increases, caught and escape

Distance metrics used	Number of trials	Number of cats	Caught percentage (%)	Evasion and escape percentage (%)
HAMA	100	4	14	86
GEODESICS			18	82
EUCLIDEAN			21	79
HAMA	100	8	30	70
GEODESICS			39	61
EUCLIDEAN			41	59
HAMA	100	12	47	53
GEODESICS			52	48
EUCLIDEAN			61	39

Table 1.
Comparative metric performances based on numerical experiments.

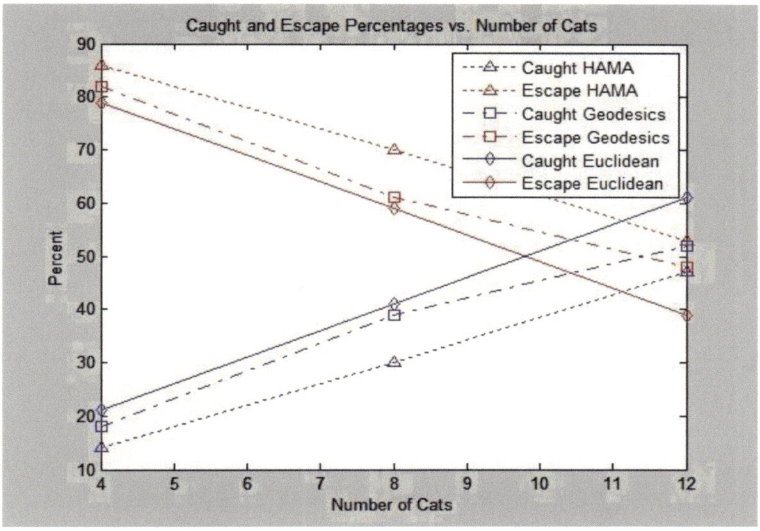

Figure 8.
The percentages of caught and escape performances of the distance metrics considered.

performances in other metrics replace. However, in this sense, compared to the increasing number of cats, HAMA still demonstrates the ability to at least maintain its initially set objective—that is, the prey can constantly escape from the pack of predators, and avert their attacks as much as possible.

The program runs the simulation until the mouse is caught, escapes the 600×600 square, or it iterates 500 times. When any of these three termination criteria is met, the solution to problem has been achieved, and thus the program will terminate its execution. Given these three termination criteria, it can never be inconsistent to define two different objective functions: they can either be formulated in such a way that, with the advantage of the heuristic metric approach, the cat travels the distance to catch the mouse in the shortest time, or the mouse easily increases the percentage of escape from the cat. The latter essentially describes the performance we would like to see and for which we have shared the results in the light of these expectations in **Table 1** above. **Figures 9** and **10** below give the

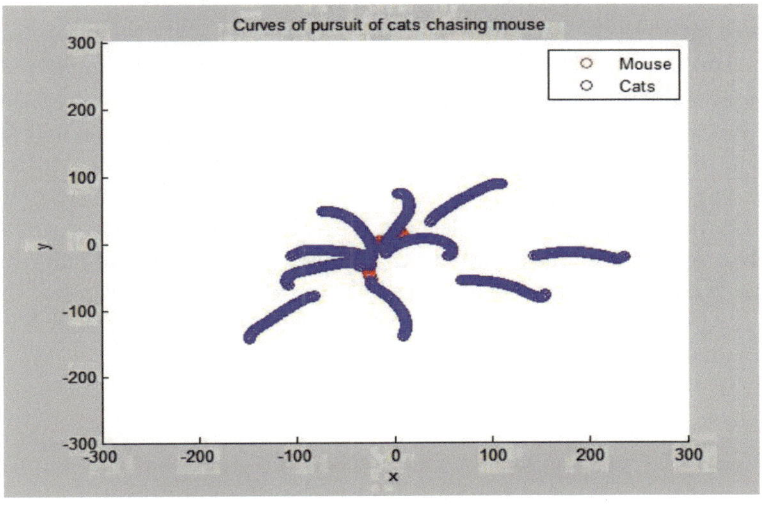

Figure 9.
Program output depicting the caught *of a single mouse for 10 randomly initialized cat states.*

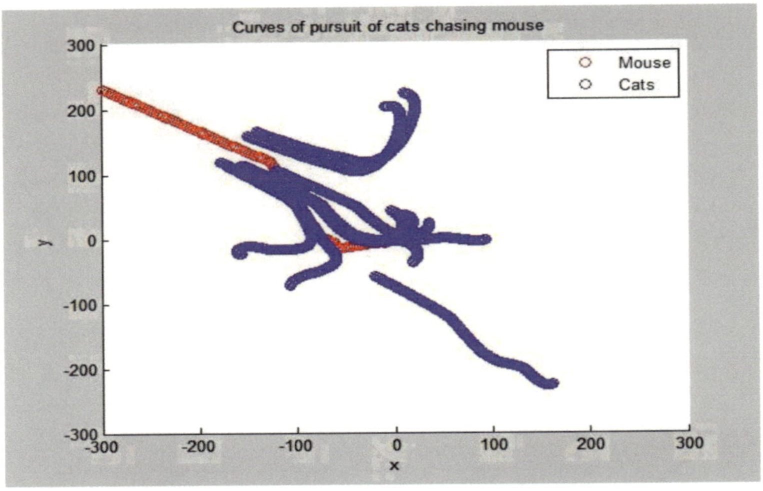

Figure 10.
Program output depicting the escape of a single mouse for 10 randomly initialized cat states.

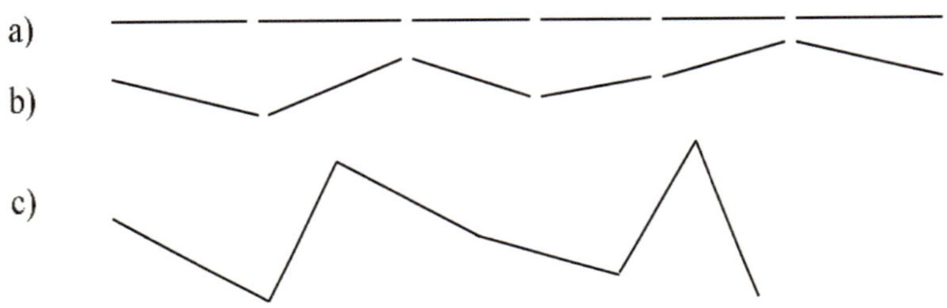

Figure 11.
Three routes of pursuit between the predator and prey, with the same metric length but with increasingly higher angular lengths [54]: a) route as simple as possible with almost zero angular variation, i.e., the shortest route, b) more complex route, but still far from random, c) the most complex route, far from any regularity, composed by angles of all amplitudes.

program output of the caught and escape problem for the case of 10 cats against a single mouse for randomly generated cat locations.

In the meantime, by emphasizing the points where we differ from different perspectives given in the literature, we find it useful to clarify the definition of angular metric that we have discussed in this work. For example, **Figure 11** shows the routes with the same metric length but with increasingly higher angular lengths [54]. The angular distance we mean in our work differs from the angular distance meant there in that: In **Figure 11(a)**, the angular distance between perfectly aligned axial segments is zero [54], whereas in our approach, while determining the heuristic metric that allows the predator to take the distance it needs to cover in order to catch its prey in the shortest time or that allows the prey to easily increase the percentage of escape from the predator, we have developed an approach that uses the angle as an effective parameter, and hence called the angular distance metric, as we explained above.

6. Conclusion

The main motivation behind our heuristic metric approach was that, consistent with the triangle inequality principle, the heuristic acqusition that started with a semi-circle and turned into a straight-line was ultimately proportional to the length of the arc seen by the angle of α of the circle-segment. The use of HAMA became even more important as the angle emerged as the most determining factor of the problem, as we explained in the relevant section above.

In prey-predator-like pursuit-evasion problems, the angles between the direction of pursuit and the direction of the line-of-sight exhibit a broad perspective corresponding to acute, right, and wide angles, depending on how the prey and the predator are positioned relative to each other for their own purposes. In the heuristic angular distance metric approach, we therefore chose to use a formula that takes into account angles rather than vector lengths.

As a distance metric application, we believe that "cat and mouse" is a good example of what is known as chase problem and the resulting trajectories are called curves of pursuit. Cat chasing a mouse problem is a very common yet interesting problem in kinematics. The trajectory, travel time and relative approach velocity of a pursuer tracking a prey along a simple curve of pursuit are deduced using basic principles of two-dimensional kinematics. Problems of this general sort are of interest to the military community and to video game designers. Here, in the context of pursuit and evasion, a design problem that allows an artificial intelligence-like process where cat and mouse makes smart moves relative to each other and therefore makes more appropriate decisions, is discussed. The design is built around Fuzzy logic control to determine route finding between the predator and prey. As long as the angle between the vectors under consideration is maintained, the major advantage of the angular distance metric is its low sensitivity to any changes in vector norms, thus providing the desired distance that is not dependent on the amplitude.

In the numerical experiments, we chose randomly generated cat locations as they better reflect the real-world in each run, and compared the results by running a total of 100 trials for each of the selected cases that accepted first 4 cats and then 8 cats and finally 12 cats against a single mouse. Thus, 'caught and escape percentages vs. number of cats' findings for three metric distances (i.e., the metric calculated only based on the Euclidean distance, the metric calculated over the geodesics definition, and the metric devised using the angle-dependent Cosine theorem as a result of our angular metric approach) have been serached for and the results evaluated comparatively. From the comparison of these three, it is clear that our approach gives better results than the others.

As for future work, the larger the graphic size representing the route of the prey and the predator and specified by the same width and height in pixels, or the higher the number of iterations, we will end up facing high computational cost of determining the minimum routes. To prevent this situation, in the near future, we are planning to consider using a subgraph around each node containing the nodes that are reachable from the origin node within the restricted distance. It may be seen as the maximum pursuer distance from the node under calculation. In this way, we will always be able to calculate the metric geodesics or the shortest paths between each pair of nodes. For example, we would use the following set of distances (in pixels): $P = \{600 \text{ (current situation)}, 1000, 1400, 2000, 4000, 8000\}$.

Conflict of interest

The author declares that there is no conflict of interest regarding the publication of this paper.

Author details

İhsan Ömür Bucak
Iğdır University, Iğdır, Turkey

*Address all correspondence to: iomur.bucak@igdir.edu.tr

References

[1] Nabiyev VV. Yapay Zekâ: İnsan-Bilgisayar Etkileşimi. Ankara: Seçkin Publishing House; 2010. p. 776

[2] Feigenbaum EA, Feldman J, editors. Computers and Thought. New York: McGraw-Hill Inc.; 1963. p. 548

[3] Burke EK, Gendreau M, Hyde M, Kendall G, Ochoa G, Ozcan E, et al. Hyper-heuristics: A survey of the state of the art. Journal of the Operational Research Society. 2013;**64**(12): 1695-1724. DOI: 10.1057/jors.2013.71

[4] Fang HL, Ross P, Corne D. A promising hybrid GA/heuristic approach for open-shop scheduling problems. In: Cohn A, editor. Eleventh European Conference on Artificial Intelligence. Chichester, UK: John Wiley & Sons; 1994. pp. 590-594

[5] Hart E, Ross P, Nelson JAD. Solving a real-world problem using an evolving heuristically driven schedule builder. Evolutionary Computing. 1998; **6**(1):61-80. DOI: 10.1162/evco.1998.6.1.61

[6] Marin-Blazquez JG, Schulenburg J. A hyper-heuristic framework with XCS: Learning to create novel problem-solving algorithms constructed from simpler algorithmic ingredients. In: IWLCS, Vol. 4399 of Lecture Notes in Computer Science. Berlin, Heidelberg: Springer; 2005. pp. 193-218. DOI: 10.1007/978-3-540-71231-2_14

[7] Burke EK, Petrovic S, Qu R. Case based heuristic selection for timetabling problems. Journal of Scheduling. 2006; **9**(2):115-132. DOI: 10.1007/s10951-006-6775-y

[8] Ross P, Marin-Belazquez JG. Constructive hyper-heuristic in class timetabling. In: IEEE Congress on Evolutionary Computation; 2–4 September 2005. pp. 1493-1500

[9] Kahng AB. Traveling salesman heuristics and embedding dimension in the Hopfield model. In: Proceedings of the IEEE/INNS International Joint Conference on Neural Networks. Washington, DC; 1989. pp. 513-520. DOI:10.1109/IJCNN.1989.118627

[10] Flach P. The geometry of ROC space: Understanding machine learning metrics through ROC isometrics. In: Proceedings of the 20th International Conference on Machine Learning (ICML-2003). 2003. pp. 194-201

[11] Janssen F, Fürnkranz J. On the quest for optimal rule learning heuristics. Machine Learning. 2010;**78**(3):343-379. DOI: 10.1007/s10994-009-5162-2

[12] Glover F. Future paths for integer programming and links to artificial intelligence. Computers and Operations Research. 1986;**13**:533-549. DOI: 10.1016/0305-0548(86)90048-1

[13] Glover F, Laguna M. Tabu Search. Boston: Kluwer; 1997. pp. 41-59. DOI: 10.1007/978-1-4419-1665-5_2

[14] Mirjalili S, Lewis A, Mostaghim S. Confidence measure: A novel metric for robust meta-heuristic optimisation algorithms. Information Sciences. 2015; **17**:114-142. DOI: 10.1016/j.ins.2015.04.010

[15] Gaschnig J. A problem similarity approach to devising heuristics: First results. In: IJCAI. 1979. pp. 301-307. DOI:10.1016/B978-0-934613-03-3.50007-6

[16] Guida G, Somalvica M. Semantics in problem representation. Information Processing Letters. 1976;**5**(5):141-145. DOI: 10.1016/0020-0190(76)90060-0

[17] Guida G, Somalvica M. A method for computing heuristics in problem solving. Information Sciences. 1979;**19**:

251-259. DOI: 10.1016/0020-0255(79)90024-0

[18] Pearl J. On the discovery and generation of certain heuristics. AI Magazine. 1983;**4**(1):23-33. DOI: 10.1609/aimag.v4i1.385

[19] Valtorta M. A result on the computational complexity of heuristic estimates for the A^* algorithm. Information Sciences. 1984;**34**:47-59. DOI: 10.1016/0020-0255(84)90009-4

[20] Passino KM, Antsaklis PJ. Metric space approach to the specification of the heuristic function for the A* algorithm. IEEE Transactions on Systems, Man, and Cybernetics. 1994; **24**(1):159-166. DOI: 10.1109/21.259697

[21] Rosenfeld A, Kaminka G, Kraus S. Adaptive robot coordination using interference metrics. In: Proceedings of the 16th Eureopean Conference on Artificial Intelligence (ECAI'2004); Valencia, Spain; 2004. pp. 910-916

[22] Bringmann K, Engels KC, Manthey B, Rao BVR. Random shortest paths: Non-Euclidian instances for metric optimization problems. Algorithmica. 2015;**73**(1):42-62. DOI: 10.1007/s00453-014-9901-9

[23] Akinyemi IO. A refinement-based heuristic method for decision making in the context of Ayo game [thesis]. Covenant University; 2012

[24] Sosa-Ascencio A, Ochoa G, Terashima-Marin H, Conant-Pablos SE. Grammar-based generation of variable-selection heuristics for constraint satisfaction problems. Genetic Programming and Evolvable Machines. 2016;**17**(2):119-144. DOI: 10.1007/s10710-015-9249-1

[25] Sosa-Ascencio A, Terashima-Marin H, Ortiz-Bayliss JC, Conant-Pablos SE. Grammar-based selection hyper-heuristics for solving irregular bin packing problems. In: Proceedings of the 2016 Genetic and Evolutionary Computation Conference Companion (GECCO'16); Denver, Colorado, USA; 2016. pp. 111-112

[26] Dokeroglu T, Cosar A. A novel multi-start hyper-heuristic algorithm on the grid for the quadratic assignment problem. Engineering Applications of Artificial Intelligence. 2016;**52**:10-25. DOI: 10.1016/j.engappai.2016.02.004

[27] Wu X, Consoli P, Minku L, Ochoa G, Yao X. An evolutionary hyper-heuristic for the software project scheduling problem. In: Handl J, Hart E, Lewis P, López-Ibáñez M, Ochoa G, Paechter B, editors. Parallel Problem Solving from Nature—PPSN XIV. Cham: Springer; 2016. pp. 37-47. DOI: 10.1007/978-3-319-45823-6_4

[28] Lozano JMC, Gimenez D, Garcia LP. Optimizing metaheuristics and hyperheuristics through multi-level parallelism on a many core system. In: IEEE International Parallel and Distributed Processing Symposium Workshops. 2016. pp. 786-795

[29] Tyasnurita R, Özcan E, John R. Learning heuristic selection using a time delay neural network for open vehicle routing. In: IEEE Congress on Evolutionary Computation (CEC). 2017. pp. 1474-1481. DOI:10.1109/CEC.2017.7969477

[30] Soria-Alcaraz JA, Ochoa G, Sotelo-Figeroa MA, Burke EK. A methodology for determining an effective subset of heuristics in selection hyper-heuristics. European Journal of Operational Research. 2017;**260**:972-983. DOI: 10.1016/j.ejor.2017.01.042

[31] Lissovoi A, Oliveto PS, Warwicker JA. Simple hyper-heuristics control the neighbourhood size of randomized local search optimally for leadingones. Evolutionary Computation. 2020;**28**(3):437-461. DOI: 10.1162/evco_a_00258

[32] Hong L, Drake JH, Woodward JR, Özcan E. A hyper-heuristic approach to automated generation of mutation operators for evolutionary programming. Applied Soft Computing. 2018;**62**:162-175. DOI: 10.1016/j.asoc.2017.10.002

[33] Schlünz EB, Bokov PM, van Vuuren JH. Multiobjective in-core nuclear fuel management optimization by means of a hyperheuristic. Swarm and Evolutionary Computation. 2018; **42**:58-76. DOI: 10.1016/j. swevo.2018.02.019

[34] Elhag A, Özcan E. Data clustering using grouping hyper-heuristics. In: Liefooghe A, López-Ibáñez M, editors. Evolutionary Computation in Combinatorial Optimization. Cham: Springer; 2018. pp. 101-115. DOI: 10.1007/978-3-319-77449-7_7

[35] Sabar NR, Yi X, Song A. A bi-objective hyper-heuristic support vector machines for big data cyber-security. IEEE Access. 2018;**6**:10421-10431. DOI: 10.1109/ACCESS.2018.2801792

[36] van der Stockt SAG, Engelbrecht AP. Analysis of selection hyper-heuristics for population-based meta-heuristics in real-valued dynamic optimization. Swarm and Evolutionary Computation. 2018;**43**:127-146. DOI: 10.1016/j.swevo.2018.03.012

[37] Silva MAL, de Souza SR, Souza MJF, de França Filho MF. Hybrid metaheuristics and multiagent systems for solving optimization problems: A review of frameworks and a comparative analysis. Applied Soft Computing. 2018;**71**:433-459. DOI: 10.1016/j.asoc.2018.06.050

[38] Choong SS, Wong LP, Lim CP. An artificial bee colony algorithm with a modified choice function for the traveling salesman problem. Swarm and Evolutionary Computation. 2019;**44**: 622-635. DOI: 10.1016/j. swevo.2018.08.004

[39] Ahmed L, Mumford C, Kheiri A. Solving urban transit route design problem using selection hyper-heuristics. European Journal of Operational Research. 2019;**274**:545-559. DOI: 10.1016/j.ejor.2018.10.022

[40] Caraffini F, Neri F, Epitropakis M. HyperSPAM: A study on hyper-heuristic coordination strategies in the continuous domain. Information Sciences. 2019;**477**:186-202. DOI: 10.1016/j.ins.2018.10.033

[41] Lin J. Backtracking search based hyper-heuristic for the flexible job-shop scheduling problem with fuzzy processing time. Engineering Applications of Artificial Intelligence. 2019;**77**:186-196. DOI: 10.1016/j. engappai.2018.10.008

[42] Babic A, Miskovic N, Vukic Z. Heuristics pool for hyper-heuristic selection during task allocation in a heterogeneous swarm of marine robots. IFAC-PapersOnLine. 2018;**51**(29): 412-417

[43] de Carvalho VR, Sichman JS. Solving real-world multi-objective engineering optimization problems with an election-based hyper-heuristic. In: International Workshop on Optimisation in Multi-agent Systems (OptMAS'18). 2018. pp. 1-15

[44] Yates WB, Keedwell EC. An analysis of heuristic subsequences for offline hyper-heuristic learning. Journal of Heuristics. 2019;**25**:399-430. DOI: 10.1007/s10732-018-09404-7

[45] Zhou Y, Yang JJ, Zheng LY. Hyper-heuristic coevolution of machine assignment and job sequencing rules for multi-objective dynamic flexible job shop scheduling. IEEE Access. 2019;**7**: 68-86. DOI: 10.1109/ACCESS.2018. 2883802

[46] Zhou Y, Yang JJ, Zheng LY. Multi-agent based hyper-heuristics for

multi-objective flexible job shop scheduling: A case study in an aero-engine blade manufacturing plant. IEEE Access. 2019;7:21147-21176. DOI: 10.1109/ACCESS.2019.2897603

[47] Oyebolu FB, Allmendinger R, Farid SS, Banke J. Dynamic scheduling of multi-product continuous biopharmaceutical facilities: A hyper-heuristic framework. Computers and Chemical Engineering. 2019;**125**:71-88. DOI: 10.1016/J.COMPCHEMENG.2019.03.002

[48] Leng L, Zhao Y, Wang Z, Zhang J, Wang W, Zhang C. A novel hyper-heuristic for the biobjective regional low-carbon location-routing problem with multiple constraints. Sustainability. 2019; **11**(6):1596. DOI: 10.3390/su11061596

[49] Lissovoi A, Oliveto PS, Warwicker JA. On the time complexity of algorithm selection hyper-heuristics for multimodal optimization. In: Proceedings of the AAAI Conference on Artificial Intelligence. 2019. pp. 2322-2329

[50] Bucak IO, Tatlilioglu M. Different viewpoint for puzzle problems as artificial intelligence toy problems: A heuristic approach. In: Book of Abstracts of the 4th International Eurasian Conference on Mathematical Sciences and Applications (IECMSA-2015); Athens, Greece; 2015. p. 273

[51] Chanwimalueng T, Mandic DP. Cosine similarity entropy: Self-correlation-based complexity analysis of dynamical systems. Entropy. 2017;**19**: 652. DOI: 10.3390/e19120652

[52] Olson E. ECE539 Semester Project: Expanded Dog Chasing Cat Problem; 2003

[53] Hu YH. Intro. ANN & Fuzzy Systems, Lectures 33, 34 and 35: Fuzzy Logic Control I, II, and III; 2001

[54] Serra M, Hillier B. Angular and metric distance in road network analysis: A nationwide correlation study. Computers, Environment and Urban Systems. 2019;**74**:194-207. DOI: 10.1016/j.compenvurbsys.2018.11.003c

Chapter 4

From ERL to MBZIRC: Development of An Aerial-Ground Robotic Team for Search and Rescue

Barbara Arbanas, Frano Petric, Ana Batinović, Marsela Polić, Ivo Vatavuk, Lovro Marković, Marko Car, Ivan Hrabar, Antun Ivanović and Stjepan Bogdan

Abstract

This chapter describes the efforts of the LARICS team in the 2019 European Robotics League (ERL) Emergency Robots and the 2020 Mohamed Bin Zayed International Robotics Challenge (MBZIRC) robotics competitions. We focus on the implementation of hardware and software modules that enable the deployment of aerial-ground robotic teams in unstructured environments for joint missions. In addition to the overall system specification, we outline the main algorithms for operation in such conditions: autonomous exploration of unknown environments and detection of objects of interest. Analysis of the results shows the success of the developed system in the competition arena of two of the largest outdoor robotics challenges. Throughout the chapter, we highlight the evolution of the robotic system based on the experience gained in the ERL competition. We conclude the chapter with key findings and additional improvement ideas to advance the state of the art in search and rescue applications of heterogeneous robotic teams.

Keywords: search and rescue robotics, aerial robotics, autonomous exploration, MBZIRC, ERL

1. Introduction

Search and rescue (SAR) problems for collaborative multi-robot systems (MRS) have been an interesting research topic for several decades [1–3]. The attractiveness of the domain stems from the variety of problems it incorporates, including mapping and situational awareness, monitoring and surveillance, establishing communication networks, or cooperative decision making. All these aspects make SAR a very difficult problem to solve. Despite the considerable difficulties, conducting SAR operations with autonomous MRS offers many advantages. A capable robotic search and rescue team can replace humans as first responders in disaster areas, map and inspect the site, and make it safe for the approach of human rescue teams.

To facilitate and accelerate progress in the field of search and rescue, many robotics competitions have been held over the years. In recent years, many of them have begun to incorporate multi-domain challenges involving robots from different domains, including aerial, ground, and underwater robots. At the forefront of this concept is the euRathlon [4], the world's first multi-domain (air, land and sea) multi-robot SAR competition. In a recreated disaster scenario inspired by the 2011 Fukushima nuclear power plant accident, the euRathlon 2015 Grand Challenge required teams of robots to work together to map the area, find missing workers and contain a leak. As a continuation of this initiative, a series of the European Robotics League Emergency Robots [5] (ERL-ER) was organized as part of the European Robotics League. Some changes were made to increase the level and difficulty of the challenges, including more complex manipulation tasks and even more collaboration between robots from different domains. Three ERL-ER competitions were held: Piombino 2017, Seville 2019 and La Spezia 2019.

In parallel, there is another branch of multi-domain competitions that gathers many teams worldwide, the Mohamed Bin Zayed International Robotics Challenge [6] (MBZIRC). MBZIRC focuses not only on SAR missions, but also on promoting the state of the art in robotics in general. The missions are very challenging from both control and algorithmic perspective. They often involve complex manipulation tasks, the need for rapid response and tracking of fast-moving objects, and SAR segments such as fire detection and extinguishing. This competition took place in 2017 and 2020 in Abu Dhabi.

In this chapter, we focus on our efforts in developing an autonomous aerial-ground search and rescue team during the last few years. Our system was formed in two robotics competitions, ERL-ER 2019 Local Tournament in Seville and MBZIRC 2020. In the first competition, we succeeded in setting up the autonomous system for search and rescue missions, including mapping the area and detecting objects of interest. The MBZIRC challenge involved a more complex scenario with three different missions, and we adapted and improved the developed system based on the results of the ERL competition. In this chapter, we detail the hardware and software design and the methodology used to solve the problems in the challenges. At the end of the chapter, we offer important conclusions and practical advice from the years of development of our robotic team to further advance the practical application of autonomous MRS to these types of problems.

The chapter is organized as follows. The next section surveys the state of the art in SAR problems for multi-robot collaborative systems. Section 3 describes the mechanical design of the robotic aerial and ground platforms. Section 4 details the software design for both platforms, including the modules used and the organization of the overall software architecture. Section 5 provides insight into the development of autonomous exploration and mapping of unknown areas using an Unmanned Aerial Vehicle (UAV) or Unmanned Ground Vehicle (UGV) with a LiDAR. In Section 6 we give an insight into the object detection methods we have developed for the ERL and MBZIRC competitions. In Section 7 we analyze the performance of the described system and its components during the competition runs, followed by conclusions and lessons learned as the last section.

2. Related work

Robotics research in the field of SAR has, to a large degree, been driven by competitions and challenges, first of which were introduced in late 90s [7], followed by later establishment of euRathlon [4], DARPA challenge [8], ERL Emergency [5] and MBZIRC [6].

Through the fundamental research in the field of robotics that is required to meet the demanding scenarios, the competitions and challenges are also driving the development of standardized benchmarks for SAR robots, with the US National Institute of Standards and Technology (NIST) leading the way [9]. NIST recognizes the urban search and rescue scenarios as a great tool to measure how intelligent autonomous robots are by looking at several components of an autonomous search and rescue robot or robot team.

The first component of a SAR system is the ability to traverse difficult terrain, which is usually solved by using a heterogeneous robotics team consisting of several robots with different capabilities. For the land-based SAR, there are various formations that are used, such as physically interacting UGV-UGV teams [10] where one robot can carry the other one, physically interacting UAV-UGV teams [11] where the UAV can carry the UGV over obstacles and the UGV can carry the UAV to save energy, or teams where only information is shared [12]. In our work for ERL and MBZIRC, we focus on the information sharing team consisting of a UGV and at least one UAV.

To traverse the terrain and find objects or people, the SAR robotic team needs to be able to perceive the surroundings. While some applications may require acoustic and thermal sensing, our robotic team is equipped with 2D and 3D LiDARs for mapping and navigation and cameras (both RGB and RGBD) for detecting and locating people or objects. For some detection tasks, we used deep neural networks [13], while some objects could be recognized solely by color, similar to [14]. 2D and 3D maps of the environment were built using Google Cartographer [15], which was adapted to use information from the GPS and to also serve as a pose sensor for UAV control [16]. The navigation stack for the UGV is based on the TEB planner [17], while the UAVs rely on the TOPP-RA approach [18].

Generating waypoints for navigation planners is an essential aspect of robot team autonomy. In our work, several methods were implemented: (a) waypoint generation using a 2D lawnmover pattern in relatively small areas of known size; (b) Levvy flight 2D waypoint generation for large areas of known size [19]; (c) autonomous 2D [20] and 3D exploration [21] for areas of unknown size. In this work, we briefly discuss our autonomous exploration approach for ERL and MBZIRC competitions, which was later extended to the planner described in [21], based on the exploration tool called 3D-FBET, work of Zhu et al. [22].

The final component of the autonomous SAR system is collaboration between multiple agents, including humans, which is constantly a topic of great interest in SAR challenges and projects [23–25]. While we have developed a framework for collaboration and coordination in heterogeneous robot teams [11, 26], due to nature of challenges in ERL and MBZIRC, the developed approaches were not deployed in the field.

3. Mechanical design of the robotic platforms

In this section, we describe the hardware setup of the robotic platforms used for the search and rescue missions in the ERL and MBZIRC competitions. We outline the mechanical structure of the robots and the components installed on board the vehicles to enable the execution of the specified tasks.

3.1 Hardware setup of the UAV

For the hardware design of the UAV, we followed the requirements of the challenge specifications. The main purpose of the aerial platform is to autonomously map and explore the disaster area and search for objects of interest. Therefore, we developed a custom hardware configuration to support this purpose, including the use of appropriate sensors for localization and vision-based algorithms. The mechanical design of the UAV was the same for the ERL and MBZIRC search and rescue missions.

The frame of the aerial platform (**Figure 1**) consists of four arms, a body, and two legs with skis, all made of carbon. The vehicle is equipped with Flame 60A 12S ESCs which are driving T-motor P60 KV170 12S motors with 22-inch carbon propellers. The autopilot used is a Pixhawk 2.0, and the maximum takeoff weight of the vehicle is 12 kg with a 2 kg payload. The UAV is equipped with an Intel NUC i7/16GB computer running Ubuntu 18.04 LTS and ROS Melodic, which processes input from ZED Mini stereo camera, Smartek Vision camera and Velodyne VLP-16 Puck Lite 3D LiDAR. Image processing for the UAV is performed using convolutional neural networks running on Intel Neural Compute Stick 2. All components of the aerial platform are powered by two LiPo 12S 14000mAh batteries, which give the vehicle a flight time of up to 30 minutes. The summary of components on board of UAV is given in **Table 1**.

Ardupilot's Copter firmware is a full-featured, open-source multicopter UAV controller. Copter is capable of handling the full range of flight requirements, from fast FPV racing to smooth aerial photography to fully autonomous complex missions that can be programmed via a range of compatible software ground stations. The entire package is designed to be safe, feature rich and open to custom applications. Copter firmware is loaded onto the Cube Black autopilot, which includes a 32-bit ARM Cortex M4 core processor with a fail-safe 32-bit coprocessor, 256 KB of RAM and 2 MB of flash memory. Integrated into the Cube are a triple redundant

Figure 1.
Front view of the aerial platform with all sensors and electronic devices on board.

Component	Specifications	Purpose
On-board computer	Intel NUC i7-8650U Quad Core 1.9GHz	Main computing unit
Autopilot	Pixhawk 2.0	Flight control, sensors (IMU, barometers)
Computing unit	Intel Neural Compute Stick 2	Neural network processing
GPS sensor	Ublox M8P GNSS reciever	Localization, navigation
LiDAR	Velodyne VLP-16 Puck Lite	Localization, mapping
Stereo camera	ZED Mini	Perception, object detection

Table 1.
UAV hardware component summary. The same setup was used in ERL and MBZIRC competitions.

intertial measurement unit (IMU) containing accelerometer, gyroscope and magnetometer as well as two barometers, which are isolated and damped.

The carrier board provides an interface for various Pixhawk-compatible peripherals, as well as power for all components of the drone. On the top of the UAV is an external Global Positioning System (GPS) sensor, whose receiver is based on the high-precision Ublox M8P GNSS module. When paired with the external GPS and antenna via a telemetry module, the vehicle's GPS (Pixhawk) can achieve centimeter-level accuracy. The copter firmware uses an Extended Kalman Filter (EKF) algorithm to estimate the vehicle's position, velocity, and angular orientation based on gyroscope, accelerometer, compass, GPS, airspeed, and barometric pressure measurements.

The Velodyne VLP-16 Puck Lite features a rotating array of sixteen infrared laser emitter-detector pairs that rotate at 300 to 1200 RPM. Each of the 903 nm lasers is fired 18080 times per second, and distances are measured based on the time of flight for each pulse. The captured distance measurements can be used to create high-quality point clouds. Each LiDAR sweep covers a vertical angle of 30 degrees with a resolution of 2 degrees and any horizontal angle with a resolution of 0.1 degrees to 0.4 degrees, depending on the RPM setting.

The ZED Mini is a stereo camera that provides high-resolution images and accurate ambient depth measurements. It is designed for the most demanding applications, such as autonomous vehicle control, mobile mapping, aerial mapping, security, and surveillance. It is actually a device with two cameras 65 mm (2.56 inches) apart, designed to mimic the human eye.

The on-board computer used in the UAV is an 8th generation Intel NUC with a powerful Intel i7-8650U Quad Core 1.9GHz processor. The computer has an integrated Intel Iris Plus Graphics 640, a 16GB Kingston SODIMM RAM module and a 250GB Kingston A2000 M.2 SSD as storage unit. The required power supply range is 12-19 V. The operating system installed on the on-board computer is Linux Ubuntu 18.04 LTS with the middleware ROS Melodic. Since the on-board computer needs to be embedded in the body of the drone and to save some weight, the outer case is removed and the computer is mounted on the bottom of the UAV.

3.2 Hardware setup of the UGV

In ground search and rescue missions, the choice of the ground mobile platform is mostly dictated by the mission and terrain requirements. In our robotic team, a Husky A200 was chosen as a platform suitable for motion in rough terrain. The competition tasks directly dictated the choice of additional hardware components. The complexity of the MBZIRC tasks was significantly higher compared to the ERL competition, and the hardware design was therefore more elaborate, as explained in this section.

3.2.1 ERL search and rescue platform

The main task of the UGV in the ERL competition was to autonomously locate objects of interest. Our search and rescue platform is shown in **Figure 2**, with the autonomous navigation sensor setup: a Sick NAV350 LiDAR, an IMU, a GPS receiver, and a real-time kinematic positioning (RTK) unit for higher precision. For reliable detection of obstacles during movement, LiDAR was mounted on top of the vehicle body, near the front end of the vehicle, so that there is no shadowing from other UGV parts. GPS and RTK units were elevated with respect to the vehicle body to avoid potential signal blockage and interference. The task detection requirements allowed the use of the Intel RealSense D-435, a commercially available RGB-D

Figure 2.
Hardware setup of the UGV and its sensors and actuators used in ERL competition.

camera, instead of more precise and expensive alternatives. Thanks to recent technological advances, the camera is small enough to be mounted in the front part of the vehicle without interfering with other sensing components, such as LiDAR beams.

Another task of the UGV was to provide first aid kits. For this purpose, a simple 1-DOF manipulator was constructed, sufficient for safe transport and easy unloading of the package. As the ERL task required delivery of two kits, two such manipulators were mounted at the front of the vehicle, outside of the camera field of view. The manipulator links were designed and 3D printed from PLA plastic (shown in red in **Figure 2**) and driven by Dynamixel servo motors.

The UGV design described was quite simple, quick to implement, and task-specific, resulting in successful mission execution. However, in the long run, it became apparent that a more complex solution would be much more beneficial for mission versatility.

3.2.2 MBZIRC mobile manipulation platform

In the MBZIRC competition, we chose to design a mobile manipulator platform, and to rely on the dexterity of the arm for execution of more complex tasks, such as repetitive pick and place in the wall construction scenario. Still, each new task requires some adaptation of the manipulator tool. We equipped the vehicle with a lightweight 6-DOF manipulator, a Schunk Powerball LWA-4P. The arm has integrated joint drives, eliminating the need for additional external control units or power converters. The setup of our platform is shown in **Figure 3**.

One of MBZIRC's challenges was a wall construction task in which the mobile manipulator had to pick up bricks from their stacks and transport them to the wall pattern, where they had to be arranged in a specified order. Again, a custom gripper was developed to magnetically grasp the ferromagnetic plates on the bricks, as shown in **Figure 4**. The gripper is lightweight, energy efficient and strong enough

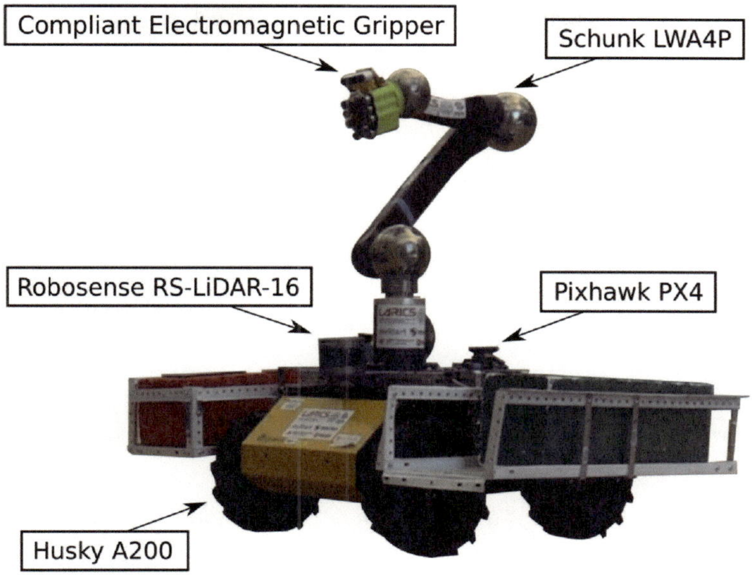

Figure 3.
Hardware setup of the UGV and its sensors, actuators, and electronic components used in MBZIRC competition.

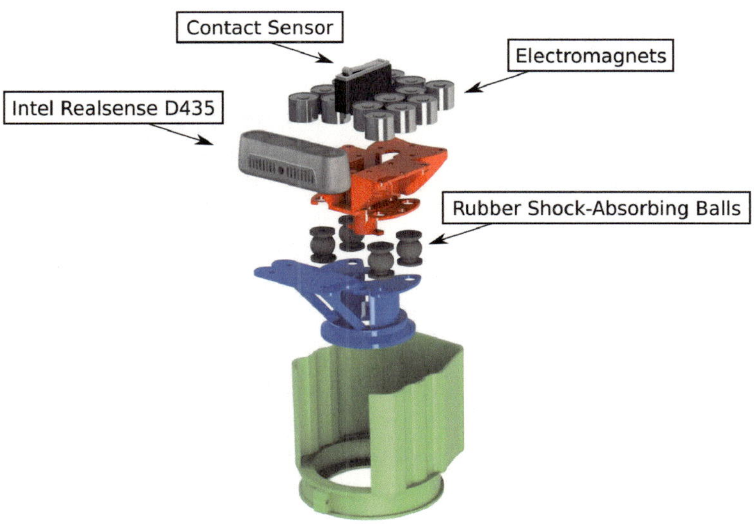

Figure 4.
Design of the passively compliant magnetic gripper for UGV brick manipulation tasks in MBZIRC challenge.

to lift up to 2 kg. The design is based on ten small electromagnets, and at least six magnets in contact are required to achieve sufficient force to lift the brick. Due to the rigidity of the mobile manipulator, any irregularity or misalignment on the contact surface could reduce the number of magnets in contact. This was solved with passive compliance, with four shock-absorbing rubber balls providing compensation for misalignment greater than 10°.

In this scenario, the RGB-D camera was used for detection, navigation, and manipulation. It was therefore mounted as an eye-in-hand on the gripper. For navigation, a 3D LiDAR (Robosense RS-LiDAR-16) was used instead of the 2D solution in ERL. As it turned out, mapping with 2D scans in realistic environments can miss some obstacles, such as those on the ground, below the beam. Other

Component	Specifications	Purpose
On-board	Intel NUC i7-8650U	Main computing unit
Computer	Quad Core 1.9GHz	
Autopilot	Pixhawk 2.0	Sensors
		(IMU, RC control)
Manipulator	ERL - two 1-DoF manipulators	Object manipulation
	MBZIRC - SchunkPowerball LWA-4P	
GPS sensor	Ublox M8P GNSS reciever	Localization, navigation
LiDAR	ERL - Sick NAV350 LiDAR	Localization, mapping
	MBZIRC - Robosense RS-LiDAR-16	
RGB-D camera	Intel RealSense D-435	Perception, object detection

Table 2.
UGV hardware component summary. The setups used in ERL and MBZIRC competitions.

hardware modifications included an additional battery pack and cargo baskets for transporting multiple bricks. The summary of the components on board the UGV, during the ERL and MBZIRC competitions, is shown in **Table 2**.

4. Software architecture and modules

Having defined the mechanical structure of the robotic platforms, in this section we address the design of the software modules implemented for each robot. Here we define the software that enables basic operations of the vehicles, such as control, localization, navigation, and trajectory planning, while challenge-specific tasks and more complex algorithms are described in the next section.

4.1 Software setup of the UAV

UAV platforms used for search and rescue missions are equipped with Ardupilot firmware. The firmware communicates with Micro Aerial Vehicle (MAV) messages using the MAVLink protocol. The Robot Operating System (ROS) is used to communicate with the on-board computer and the drone. The components used in the ROS environment are as follows:

- **MavRos**[1] is the most important component of the UAV software setup as it provides two-way communication between the vehicle firmware and the on-board computer. Its main task is to translate MAV to ROS messages and vice versa.

- **On-board UAV Controller** receives trajectory points and the current UAV state as input and produces a control output for the vehicle consisting of the desired rotation and thrust. The underlying implementation is a cascade of Proportional-Integral-Derivative (PID) controllers. The outer loop computes the velocity command based on the local position measurements, while the

[1] MAVLink extensible communication node for ROS with proxy for Ground Control Station. http://wiki.ros.org/mavros

inner loop calculates the desired orientation and thrust based on the velocity measurements.

- **UAV State Machine** serves as a position command multiplexer. It either computes commands based on the Radio Controlled (RC) controller input or forwards an external command. It also implements simple takeoff and landing procedures. The states are as follows:
    - HOLD GROUND: the UAV is on the ground and ready to arm and takeoff.
    - HOLD AIR: the UAV is in the air and ready to receive external commands.
    - JOY AIR: the UAV is in the air and is controlled by RC.
- **Trajectory Planner** receives a set of waypoints and outputs trajectory points that are sent to the on-board controller. The underlying implementation is a Time-Optimal Path Parametrization by Reachability Analysis (TOPP-RA) [18] which provides an optimal trajectory given the specified kinematic constraints.
- **Cartographer** is a method for Simultaneous Localization and Mapping (SLAM) used to determine the UAV pose in the map frame. Combined with a constant velocity model, the linear Kalman filter yields estimated UAV odometry that is used as feedback for the on-board UAV controller.

In **Figure 5** we have illustrated the interaction of the above components within the UAV software architecture on a NUC computer. This figure refers to the case where the UAV state machine is in the HOLD AIR state. The pipeline begins with inputs from the LiDAR and IMU that are fed into the Cartographer SLAM

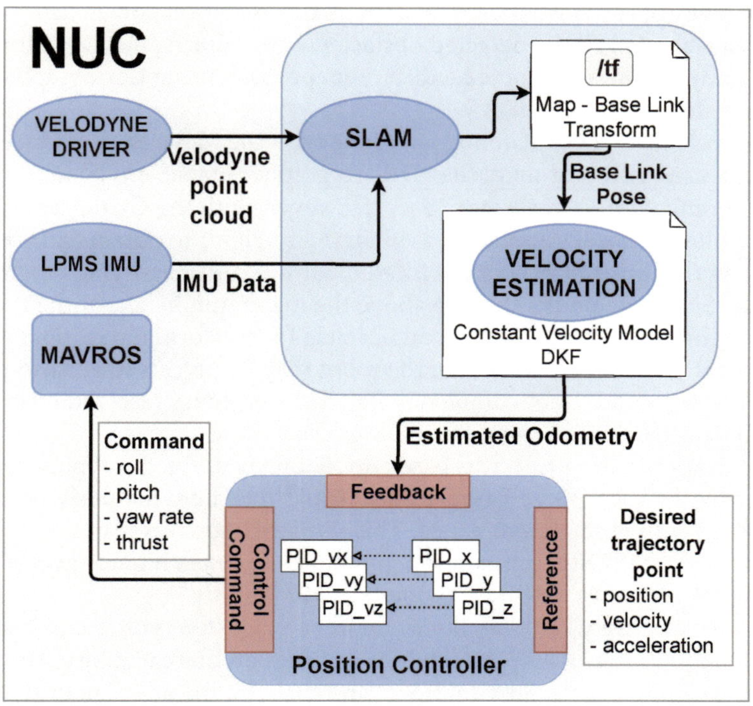

Figure 5.
Layout and the interaction of the UAV software components running on the on-board NUC computer.

algorithm, which creates the map of the environment and locates the vehicle in the map frame. Next, using the Constant Velocity Model, the estimated odometry of the vehicle is calculated and fed to the UAV position controller. Based on the reference generated with the trajectory planner, the controller generates control outputs of the desired vehicle orientation and thrust and sends them to the MavRos. These messages are then forwarded to the lower level UAV controllers implemented in the firmware. If georeferencing is needed, GPS measurements can also be forwarded to the Cartographer.

4.2 Software setup of the UGV

The main software modules used for UGV control in the competition search and rescue missions are those for localisation, motion and manipulation:

- **Cartographer SLAM** is the module used for simultaneous mapping and localization of the vehicle.
- **Move Base** is a ROS package used for ground vehicle navigation.
- **Manipulation module** for controlling two simple 1-DoF manipulators used to drop aid packages.[2]

The Google Cartographer package was used to create a map of the unknown environment and localize the mobile base within it. During the ERL-ER competition, a two-dimensional version of the Cartographer's SLAM algorithm was used, utilizing 2D laser scan data from the Sick NAV350 LiDAR. This approach results in only detecting obstacles that are aligned with the LiDAR, and thus led to problems with lower obstacles, more specifically the lower part of the obstacles as shown in **Figure 14**. The approach we took to overcome this issue was solved using a large safety margin around all the detected obstacles. Later, during our work for the MBZIRC competition, we adopted a different approach using Cartographer's 3D SLAM algorithm, and a 3D LiDAR.

Ground vehicle motion planning was done with the Move Base package. Since this package uses a 2D cost map, the 2D Cartographer map could be used for safe motion planning with obstacle avoidance. However, with the Cartographer's 3D SLAM, we filtered the 3D submaps based on their height, to obtain a 2D cost map. Points below the lower threshold are filtered out to avoid classifying the ground as an obstacle. Similarly, points that are above the maximum height of the robot are removed as they do not represent a real obstacle for platform navigation. However, ground points that are far away from the robot may be perceived as false-positive obstacles if the ground is not completely flat. This issue was resolved by creating newer submaps, generated using the 30 most recent laser scans.

The navigation during ERL was based on manually defined waypoints, selected in the map by the user. Move Base package would then autonomously perform the path planning with obstacle avoidance. This semi-autonomous planning system was improved for the MBZIRC competition, by implementing a higher-level planning algorithm that generates desired waypoints autonomously.

Thanks to the Husky A200 ability to rotate in place, the most commonly used navigation planner the DWAP, which heavily relies on this capability. However, such motion causes significant vibrations, and the with the addition of the robotic

[2] Used only in ERL competition.

arm for the MBZIRC competition, this problem is further exacerbated due to the increase in the overall mass and heightening of the center of mass position. Therefore, in the MBZIRC competition we switched to the Timed-Elastic Band (TEB) planner [17] for car-like robots, which uses online trajectory optimization to create a plan constrained by the minimum turning radius.

5. Evolution of the autonomous exploration system

Exploration and mapping of unknown environments is a fundamental task in robotics. It can be used in many different applications such as inspection, surveillance, 3D reconstruction, and search and rescue. Searching for an object of interest in an unknown environment can be formulated as an iterative process consisting of a map update, selection of a next goal, and navigation to that target. The process is complete when the object of interest is found.

Analogously, the search strategy in the exploration of an unknown environment is similar, but with different objectives. In exploration, the main goal is to create a map of the environment, while in search, the focus is on finding the object of interest. Although search is a central task in many search and rescue scenarios, we use an exploration strategy adapted for the search application.

Typical exploration methods are based on frontiers [27] and are used in both 2D and 3D space. In contrast to 2D exploration and mapping strategies, mapping large environments in 3D requires a considerable amount of memory and computational effort.

To create a map of the environment and locate a robot in it, we use a submap-based graph SLAM method Google Cartographer [15]. The map consists of submaps created from a sequence of sensor scans through scan matching, fusion with IMU and odometry. The map is then used for exploration and robot navigation.

5.1 From manual to autonomous exploration

In the ERL competition, some of the objectives were to locate missing workers, entrances to the building, and damage to structures while creating a map of the environment. The proposed layout for the ERL competition is an area of approximately 200 m by 30 m with obstacles. Since the ERL arena may contain GNSS - degraded areas, we used a SLAM algorithm for map generation and robot localisation. We adopt Cartographer SLAM to generate a 3D map onboard the robot in real time. Objects of interest are detected during semi-autonomous navigation, where an operator manually selects desired waypoints for the robot to reach autonomously. Both 2D and 3D maps of the environment are generated, and the positions of objects of interest are marked on the map.

In the MBZIRC challenge, we consider autonomous exploration and search of a UAV in an unstructured outdoor environment. In this challenge, we used only one UAV because it can fly over the area quickly and has the advantage of bird's eye view. The goal is to explore and map the environment while searching for the object of interest (a fire). Initially, the area is unknown, while it is assumed that the boundaries of the area are known. Since the area for exploration and mapping is limited, the exploration is finished when the entire area is covered or when the objects of interest are found.

In order to explore an area, create a map and find the objects of interest, the search strategy is formulated. The search strategy algorithm generates waypoints for trajectory planning considering the boundaries of the environment. In addition,

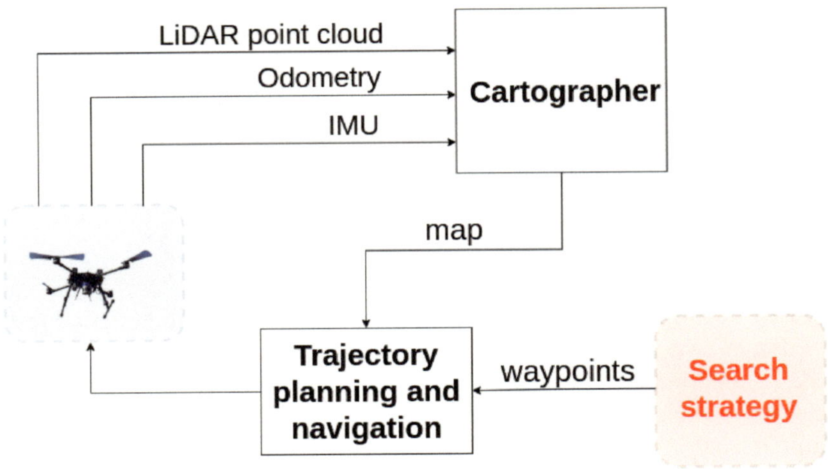

Figure 6.
Overall schematic diagram of autonomous exploration and search. The cartographer SLAM creates a map, that is used for the trajectory planning module. The generated trajectory navigates the robot towards the waypoint.

the strategy aims to explore the environment as quickly as possible while trying not to miss the objects of interest.

Waypoints can be generated in a variety of forms. For our scenario, we use the lawnmower pattern with the predefined horizontal and vertical size of an area and with the spacing value indicating how far apart the segments of the trajectory are. The lawnmower pattern is easy to implement and is suitable for exploration and search for the desired object.

The search strategy represents an input to the trajectory planning module, which generates the trajectory from the given waypoints. Objects of interest are detected during robot navigation. An overview of the proposed system is shown in **Figure 6**. The map provided by Cartographer is reliable and enables successful exploration while searching for fire.

In autonomous exploration, a suitable trajectory planner should be used to navigate the robot to the desired point. After the search algorithm computes the points in the specified arena, they are forwarded as waypoints to a trajectory planner. The robot starts following the planned trajectory and navigates to the next point. At the same time, the algorithm for detecting objects of interest runs in the background. For the safe navigation of the robot, we use the trajectory following solution described in [11].

6. Detection of objects of interest

Another major feature of the SAR systems is the ability to detect various objects of interest. In the ERL, the robots had to detect four different objects – a blocked entrance (blue rectangle), an unblocked entrance (green rectangle), damage to the building (red rectangles), and a missing worker (a dummy wearing a helmet and a safety vest). Some of the aforementioned objects are shown in the ERL arena in **Figure 7**. In the MBZIRC competition, one of the challenges was to locate and extinguish fires. Although there were variants of fires to be extinguished with water, we focused on the ground fires over which the robots had to throw blankets, as shown in **Figure 8**.

We used different strategies to detect objects of interest in these two competitions. For the ERL competition, we chose to use visual recognition using a neural

Figure 7.
Objects of interest to be detected in ERL competition. Images display the missing worker mannequin, obstructed entrance (blue rectangle), and damages to the structures (red rectangles).

Figure 8.
Ground fires to be found and covered with blankets during MBZIRC competition (Image source: https://www.mbzirc.com/photo-album).

network, for which a custom training dataset had to be created, even though the exact object was not known a-priori. To mitigate these issues and simplify the approach, we decided to use a color segmentation algorithm for object detection in the MBZIRC challenge. The simplicity of the objects to be found in this semi-structured challenge and the straightforwardness of the method and its application were the main reasons for this decision. Both methods and their properties are described in the rest of this section.

6.1 Neural network for ERL object of interest detection

The ERL search and rescue mission required recognition of several simple markers, and a human figure wearing a helmet and a vest, representing a missing worker. Even though the markers were rectangular and easily distinguished by color, we opted for the deep learning-based approach to detection because of the complexity of human figure recognition. For this reason, we relied on the transfer learning approach, taking advantage of the pre-trained deep convolutional network based on the YOLO architecture. Since we only needed to recognize 4 object classes, but needed this in real-time on constrained computational resources, we opted for the tiny-yolo version of the network (**Figure 9**).

For our detection task, we created a small custom training dataset. A large number of images with simple rectangular objects can easily be created in different environments, with the limiting factor in the cost of manual labeling procedure. For the case of a mannequin representing the missing worker, the difficulty was in not knowing exactly what this object would look like. The final dataset consisted of around 150 images, out of which around 50 contained the human figure various poses and cluttered environments, and around 100 contained the rectangular markers.

Finally, we trained the network using DarkNet implementation, with the following parameters: $batch = 64$, $subdivision = 16$, $max\_batches = 500200$ on a 4GB Nvidia GeForce GTX 980. To prevent overtraining, we saved the trained weights every 1000 iterations of learning, and in the end ran a forward pass on the test set for all of the saved stages. We chose the network weights based on the mAP score on the test set.

6.2 Color-based image filtering for object detection

The detection tasks in the MBZIRC competition were simpler compared to ERL, allowing for solution based on color segmentation. Hence, the fire detection in the third challenge of the MBZIRC competition was performed using color-based image filtering (**Figure 15**). The image received from the camera is filtered using pre-tuned Hue-Saturation-Value (HSV) thresholds. Contour detection is then performed on the filtered image, and contours with areas below a lower threshold are disregarded. If multiple contours remain, the one with the largest area is selected as the detected fire. The center of the contour is used as its position for the visual servoing procedure.

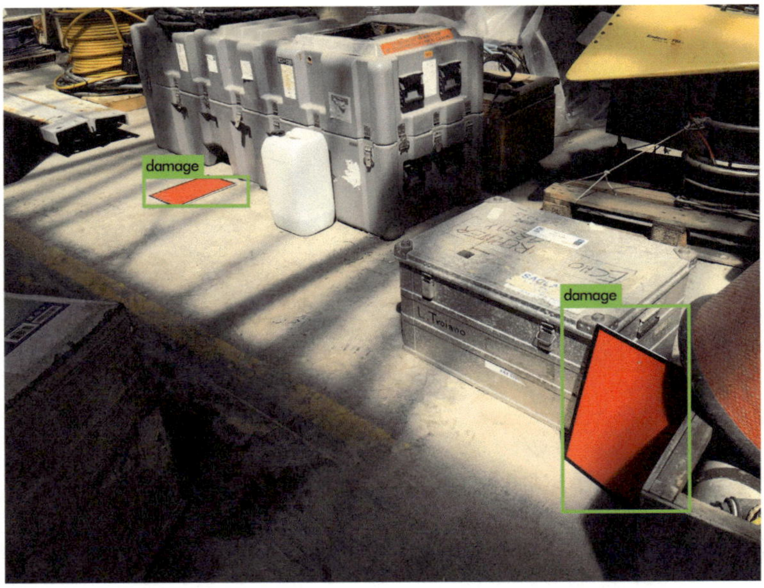

Figure 9.
Detection of the red rectangles using YOLOv3 real time object detection.

6.2.1 MBZIRC brick detection

Although this chapter focuses on SAR applications of MRS, one of the challenges during the MBZIRC competition was to achieve a more precise object manipulation in the wall building scenario that could be easily applied to the search and rescue domain. Our proposed solution involved precise 3D object detection that can be applied to other objects that need to be handled during emergency operations.

In the competition, the robots had to autonomously manipulate brick-shaped objects to build a wall. A ferromagnetic patch was attached to the top of the bricks for grasping. At the beginning of the challenge, the bricks were arranged as structured stacks, separated by color.

Initially, stacks of different colors are detected using a method similar to that used for fire detection, with different HSV thresholds used for red, green, blue and orange brick stacks. To distinguish individual bricks, ferromagnetic patches are detected within the convex hull of a detected brick stack. Contours resembling a rectangle are selected as candidate objects. To select a single brick for manipulation, a scoring scheme is used to calculate the desirability of each candidate. The desirability is calculated based on the area and image position of the candidate contour:

$$S = -w_x|x_p| + w_y y_p + w_A A \tag{1}$$

where A is the candidate area, and w_x, w_y and w_A are non-negative weights used to tune the scoring scheme and correspond to the x and y position in the image and the area, respectively. Switching from one patch candidate to another occurs only when the score of the new candidate is larger than that of the current one by some margin, to prevent rapid switching between candidates.

Our approach to determining the real-world pose of the detected magnetic patch, was to use an inverse of the Perspective-N-Point problem [28]. Assuming that the ferromagnetic patch is parallel to the ground, the positions of the endpoints of the detected rectangle in the image can be transformed into real positions using:

- Real-world camera position - from localization and manipulator kinematics.
- Height of the ferromagnetic patch - from the organized pointcloud obtained from the RealSense camera.
- Known width of the ferromagnetic patch.
- Camera calibration parameters.

7. Experimental results and evaluation

In this section, we present the results obtained during the challenge runs in the two robotics competitions. We analyze the performance of our robotic team in the autonomous mapping and exploration and object detection tasks.

7.1 Competition performance

Looking from the lens of the robotic competitions, we first present the results of the proposed UAV-UGV team in the competition arena. Here we provide challenge descriptions, and the scores obtained during the trials in the ERL-ER and MBZIRC.

7.1.1 ERL-ER competition score

The target scenario of ERL-ER is emergency response in an industrial environment affected by an earthquake. The UAV-UGV team serves as a first responder in this disaster-stricken area to secure the perimeter and enable the arrival of human rescue teams. The robots must search for two missing workers, find them as quickly as possible, and provide both with an emergency kit. The robots must also check the condition of the building after the earthquake. To do this, they output a detailed map of the surroundings to assess the safety of the area.

The competition took place in an outdoor arena of approximately $200m \times 30m$ with obstacles that the robots had to detect and navigate around. The competition arena also included a marquee that the aerial and ground robots had to enter. Rectangular red, green and blue A3-sized objects representing damage, blocked entrances and unobstructed entrances were distributed throughout the field.

At the tournament in Seville, five teams competed in a four-day competition. As specified in the rules of the competition, each team's final score was determined as the median value of all four attempts. **Table 3** shows the full results of all teams and their attempts, as well as the final score. At the expense of a lower score, we aimed for full autonomy during all trials. Therefore, the UAV, which was in an earlier stage of development, did not perform as well. However, the UGV did great, and we were able to use it to test the SLAM algorithm, as well as object detection. This tournament proved to be an excellent development and testing ground for the upcoming MBZIRC competition.

Other teams took the manual approach, such as teleoperation. For example, the user interface of Raptors was built in LabVIEW and the communication layer is based on the ROS software [30]. The system they developed supports the operator during teleoperation and during partial autonomy of the robots.

7.1.2 MBZIRC competition score

The MBZIRC competition involved three different challenges – tracking and capturing a target, building a wall, and disaster response in a fire scenario. Here we focus only on the latter, as it falls within the scope of SAR missions.

Challenge 3 of MBZIRC comprised a team of UAVs and a UGV working together to autonomously extinguish a series of simulated fires in an urban high-rise firefighting scenario. The arena, measuring $50m$ by $60m$, contained a tall structure (up to $20m$ high) simulating a high-rise building. Fires were simulated at various random locations on the floor of the arena (interior and exterior) and at various heights (ranging from $5m$ to $18m$) of the building to simulate a firefighting scenario

	Wed, Feb 20		**Thu, Feb 21**		**Fri, Feb 22**		**Sat, Feb 23**		
TEAM	**A**	**PB**	**A**	**PB**	**A**	**PB**	**A**	**PB**	**median**
LARICS	**7**	**0**	**19**	**2**	**19**	**0**	**24**	**1**	**19**
Raptors	0*	0	16	0	20	0	27	3	18
ENSTA Bretagne	0*	0	9	1	17	1	25	2	13
KAUST	0*	0	8	1	11	1	18	0	9.5
ENSTA Paris	0*	0	9	2	0*	1	19	1	4.5

Table 3.
*Team results by day and the total score (median) for the ERL-RL Seville 2019 competition [29]. Column A represents team scores, PB penalized behaviors, while attempts marked with * had disqualifying behaviors (operating outside allowed limits or no map provided).*

in a high-rise building. Teams were scored based on the number of tasks completed (fires extinguished), locations of fires extinguished, type of task completion (UAV, UGV, fire extinguishers, fire blankets), and speed of completion.

Our strategy for the challenge was to primarily focus on the ground fires, because they required minimal changes to the robot design. Therefore, we attempted to detect and extinguish the outdoor fires using a fire blanket deployed by the UAV. During the two trials, we managed to partially cover one of the two existing ground fires. This attempt was enough to place us in seventh place out of 20 participating teams. Although seemingly simple, the task proved very challenging for many teams, as shown by the results presented in **Table 4**.

An approach similar to ours was used by a team consisting Polytechnical University of Madrid, University Pablo Olavide and Poznan University of Technology [32]. In their localization, they also used a map-based approach and used 3D LiDAR scans in LOAM: Lidar Odometry and Mapping algorithm. A different approach was taken by the team from Czech Technical University in Prague, University of Pennsylvania and NYU [33] who used Global Navigation Satellite System (GNSS) for outdoor localization and a 2D LiDAR and stereo camera for indoor localization. For fire detection, both teams decided to use infrared cameras to detect heat sources. In the case of wall fires, this was a logical solution. Since we were not targeting these fires, our image-based detection worked just as well for the red floor fires. These results demonstrate the difficulty of SAR applications in real-world conditions in outdoor arenas and the potential for great progress in this area of research.

7.2 Autonomous mapping and exploration

In the field of autonomous mapping and exploration, we can observe an evolution of approaches. In the ERL competition, we explored the area manually (using an RC controller) or semi-autonomously (generating waypoints in the graphical interface and reaching them autonomously). For the MBZIRC challenge, we extended the method to autonomous exploration, where the robots could search the area and expand the map based on the autonomous exploration algorithm described in Section 5.

TEAM	Score	Time left	Rank
University of Seville, Tecnico Lisboa (IST Lisbon), CATEC	12.2625	0	1
Technical University of Denmark	10	0	2
University of New South Wales Sydney	10	0	2
Czech Technical University in Prague, University of Pennsylvania, NYU	7	0	4
Korea Advanced Institute of Science and Technology (KAIST)	7	0	4
University of Tokyo	5	120	6
UNIZG Faculty of Electrical Engineering and Computing	**5**	**0**	**7**
Polytechnic University of Madrid, University Pablo Olavide, Poznan University of Technology	5	0	7
Virginia Tech	4.5	0	9

Table 4.
The results of the best 9 out of 20 participating teams in the MBZIRC 2020 competition. These results were obtained in fully autonomous mode and therefore rank above manual approaches. The full results are available at [31].

The maps created during the ERL competition with UGV are shown in **Figure 10**, and 2D and 3D UAV maps are shown in **Figure 11**. In both the 2D and 3D maps, we see high-resolution detailed maps of the competition arena. Here, we demonstrate the ability of the proposed system to generate accurate maps during the mission runtime. However, in MBZIRC, we have gone a step further and combined the map generation with the autonomous waypoint selection method. The result of the obtained map can be seen in **Figure 12**.

The complete trajectory performed by an UAV is shown in **Figure 13**. The trajectory is obtained during the first experiment where we used manual navigation. The drift after loop closure was less than 0.3 m. We can conclude that Cartographer SLAM is useful when we navigate in the map and when we have a closed loop with trajectory planning and navigation. A more detailed analysis of Cartographer performance can be found in [16].

It can be observed that the accuracy of the final maps is not perfect compared to the mobile mapping technologies. This is due to the use of Cartographer SLAM, as it requires a feature-rich environment, a well-sampled IMU and loop closure to generate as accurate map as possible. In [16], it is shown that Cartographer accumulates more drift than other state-of-the-art strategies, but Cartographer can detect loop closure independently and the drift can be corrected. In other words, Cartographer accumulates a significant amount of drift (up to 2 meters), but upon returning near

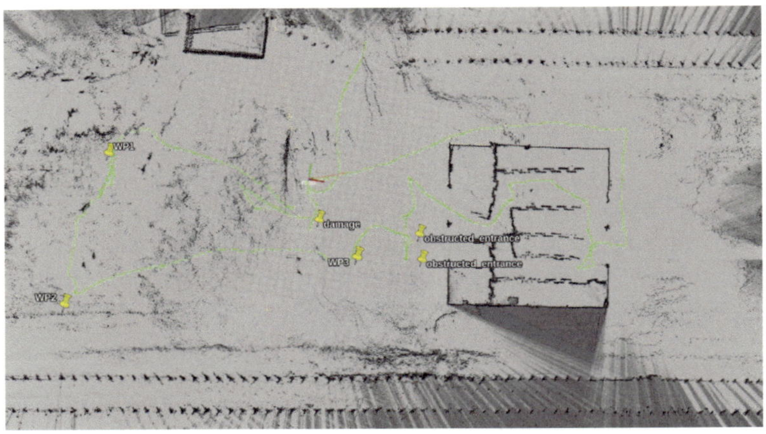

Figure 10.
2D map built by UGV during an ERL competition run.

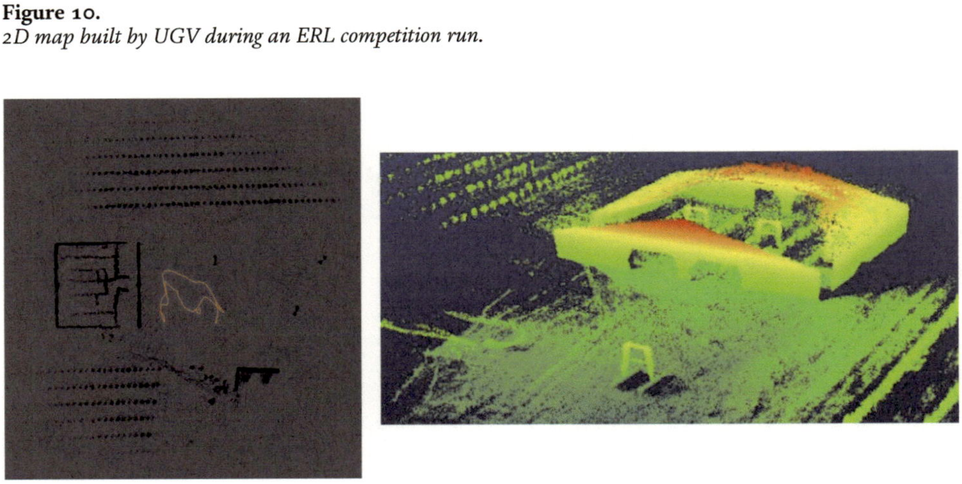

Figure 11.
2D (left) and 3D (right) map of the outdoor and indoor area of ERL arena generated by the UAV.

DOI: http://dx.doi.org/10.5772/intechopen.99210

the start position, this drift is corrected (up to 0.2 meters) by closing the loop through the global SLAM.

In addition, closing the loop allows the correction of the cumulative drift, which provides higher pose and map accuracy during the flights. Due to the specific tasks performed in the challenges, closing the loop is not guaranteed to provide an accurate map. Future work will include significant efforts to improve map accuracy so that the map can be used not only in autonomous search and exploration tasks, but also in tasks that require a more accurate map, such as autonomous wall building, obstacle avoidance, and building inspection.

Figure 12.
Side view (left) and top view (right) of the 3D map of the outdoor MBZIRC area created with the UAV.

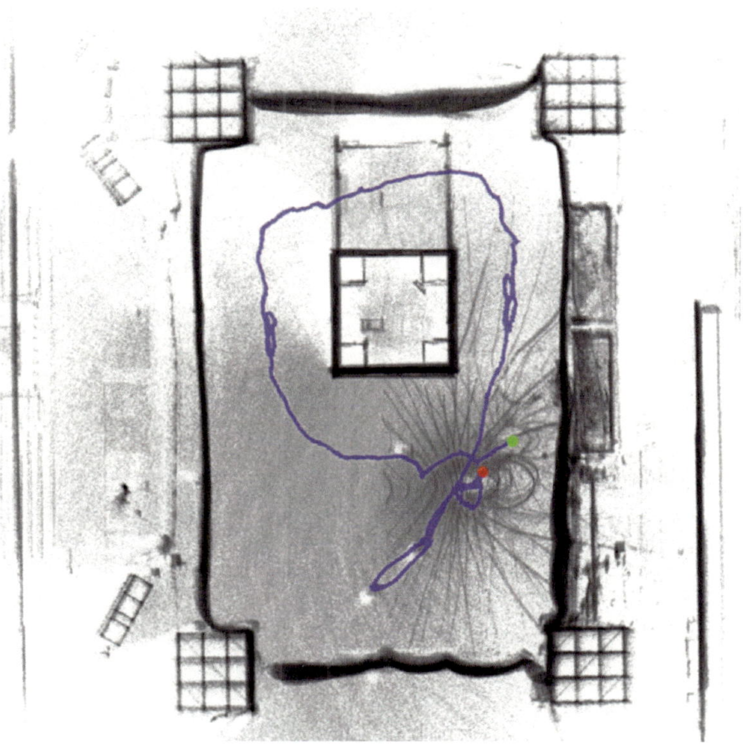

Figure 13.
A trajectory performed by an UAV in the MBZIRC challenge.

7.3 Detection of objects of interest

Next, we analyze the performance of the two object detection methods described in Section 6. In **Figure 14** we show several examples of detection of objects of interest by UGV using the algorithm described in 6 for ERL competition. These images were taken by a UGV on-board camera during several mission runs. The positions of the first three detected objects match those on the UGV map in **Figure 10**. We were able to detect the majority of the searched objects during the mission run. This mainly refers to the rectangular objects indicating blocked and unblocked entrances as well as damage to the building. However, due to the specifics of the missing worker mannequin, our neural network training set did not include the exact figure in any of the images. Therefore, we had difficulty performing this detection. This could be mitigated by training the network with a dataset containing a more accurate representation of the object being searched for, or by using a different approach. For example, color-based image segmentation might be appropriate since the safety vest has a very distinct orange color.

The output of the color-based fire detection algorithm in the MBZIRC competition is shown in **Figure 15**. The left side shows the original image, while the right side shows the image masked based on the predefined HSV threshold values. The green outline represents the object contour, while the yellow circle in the center represents the center of the object in the image. This approach provides a fast and efficient method for detecting simple objects of interest, which is certainly the case

Figure 14.
Image frames showing detections of objects of interest during ERL competition runs of the UGV.

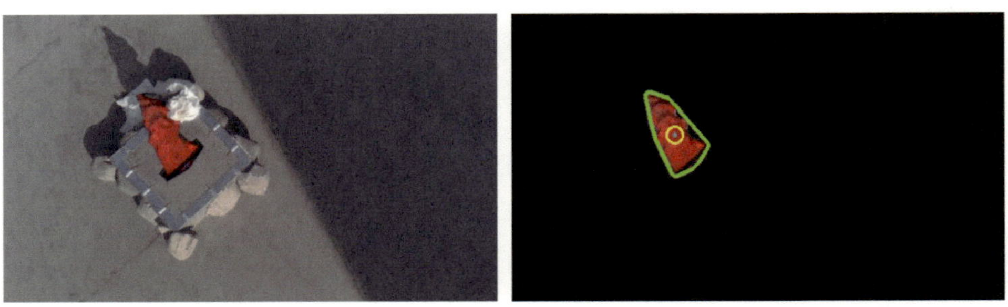

Figure 15.
Fire detection for the third MBZIRC challenge.

with a bright red fire. However, some care must be taken when setting the HSV thresholds, as the image colors are dependent on the illumination properties, which can change depending on outdoor lighting conditions.

8. Conclusions and lessons learned

This chapter presented the development of an autonomous aerial-ground search and rescue robotic team at the University of Zagreb, driven mainly by two large competitions: European Robotics League and Mohamed Bin Zayed International Robotics Challenge.

The hardware aspect of the robot team is developed considering specific requirements of the challenges, with the challenge-specific design decisions being more pronounced in our UGV design, ranging from simple 1DOF ejector arms to 6DOF mobile manipulator. On the other hand, our aerial platform did not require major hardware changes, as challenges did not require the UAV to interact with the environment on a similar scale to UGVs, but the trend is clearly changing and it can be expected that future challenges will require more and more interaction with the environment from the UAVs.

The biggest lesson to be learned from competition driven development, as evidenced by the software stack described in this chapter, is to design a modular system, both on the level of a single robot (interchangeable sensors and actuators that can easily be plugged into the existing control structure) but also on the level of a team. The modularity of our SAR robotic team is shown in using different sensors and algorithms for UGV mapping and navigation, and easily extending exploration strategies for the UAV.

Given the current state of our system, and the competitions being expected to become more and more challenging, the navigation through the unknown difficult environment is of utmost priority, which will require semantic segmentation of built 3D maps to differentiate traversable terrain from obstacles, both for the UGV and UAV. With the use of UAVs in SAR going from large open spaces to more confined urban areas, it is necessary to further develop the mapping and localization algorithms to provide a stable localization feedback for control and trajectory execution of UAVs in obstacle-rich environments.

Acknowledgements

This work has been supported by European Commission Horizon 2020 Programme through project under G. A. number 810321, named Twinning coordination action for spreading excellence in Aerial Robotics - AeRoTwin and by the Mohammed Bin Zayed International Robotics Challenge. The work of doctoral students Barbara Arbanas (DOK-2018-01), Ana Batinović (DOK-2018-09), Marko Car (DOK-2015-10), and Marsela Polić (DOK-2018-09) has been supported in part by the "Young researchers' career development project–training of doctoral students" of the Croatian Science Foundation. DOK-2018-01 is financed by the European Union from the European Social Fund.

Abbreviations

ERL	European Robotics League
MBZIRC	Mohamed Bin Zayed International Robotics Challenge

UAV	Unmanned Aerial Vehicle
UGV	Unmanned Ground Vehicle
SAR	Search and Rescue
MRS	Multi-Robot Systems
RTK	Real-Time Kinematic positioning
TOPP-RA	Time-Optimal Path Parametrization by Reachability Analysis
SLAM	Simultaneous Localization and Mapping

Author details

Barbara Arbanas*, Frano Petric, Ana Batinović, Marsela Polić, Ivo Vatavuk, Lovro Marković, Marko Car, Ivan Hrabar, Antun Ivanović and Stjepan Bogdan
Laboratory for Robotics and Intelligent Control Systems, Faculty of Electrical Engineering and Computing, University of Zagreb, Zagreb, Croatia

*Address all correspondence to: barbara.arbanas@fer.hr

References

[1] J. P. Queralta, J. Taipalmaa, B. Can Pullinen, V. K. Sarker, T. Nguyen Gia, H. Tenhunen, M. Gabbouj, J. Raitoharju, and T. Westerlund, "Collaborative multi-robot search and rescue: Planning, coordination, perception, and active vision," *IEEE Access*, vol. 8, pp. 191617–191643, 2020.

[2] D. Drew, "Multi-agent systems for search and rescue applications," *Current Robotics Reports*, vol. 2, pp. 189–200, 06 2021.

[3] J. Delmerico, S. Mintchev, A. Giusti, B. Gromov, K. Melo, T. Horvat, C. Cadena, M. Hutter, A. Ijspeert, D. Floreano, L. M. Gambardella, R. Siegwart, and D. Scaramuzza, "The current state and future outlook of rescue robotics," *Journal of Field Robotics*, vol. 36, no. 7, pp. 1171–1191, 2019.

[4] A. F. T. Winfield, M. P. Franco, B. Brueggemann, A. Castro, M. C. Limon, G. Ferri, F. Ferreira, X. Liu, Y. Petillot, J. Roning, F. Schneider, E. Stengler, D. Sosa, and A. Viguria, "euRathlon 2015: A multi-domain multi-robot grand challenge for search and rescue robots," in *Towards Autonomous Robotic Systems*, pp. 351–363, Springer International Publishing, 2016.

[5] A. F. T. Winfield, M. Palau Franco, B. Brueggemann, A. Castro, G. Ferri, F. Ferreira, X. Liu, Y. Petillot, J. Roning, F. Schneider, E. Stengler, D. Sosa, and A. Viguria, "euRathlon and ERL Emergency: A multi-domain multi-robot grand challenge for search and rescue robots," in *ROBOT 2017: Third Iberian Robotics Conference*, pp. 263–271, Springer International Publishing, 2018.

[6] "MBZIRC competition." https://www.mbzirc.com/. Accessed: 2021-05-27.

[7] H. Kitano, S. Tadokoro, I. Noda, H. Matsubara, T. Takahashi, A. Shinjou, and S. Shimada, "Robocup rescue: search and rescue in large-scale disasters as a domain for autonomous agents research," in *IEEE SMC'99 Conference Proceedings. 1999 IEEE International Conference on Systems, Man, and Cybernetics (Cat. No.99CH37028)*, vol. 6, pp. 739–743 vol.6, 1999.

[8] "DARPA robotics challenge." https://www.darpa.mil/program/darpa-robotics-challenge. Accessed: 2021-05-27.

[9] A. Jacoff, E. Messina, and J. Evans, "A standard test course for urban search and rescue robots," Performance Metrics for Intelligent Systems, Workshop — — — NIST, 2000-08-01 2000.

[10] G. D. Cubber, D. Doroftei, K. Rudin, K. Berns, A. Matos, D. Serrano, J. M. Sanchez, S. Govindaraj, J. Bedkowski, R. Roda, E. Silva, S. Ourevitch, R. Wagemans, V. Lobo, G. Cardoso, K. Chintamani, J. Gancet, P. Stupler, A. Nezhadfard, M. Tosa, H. Balta, J. Almeida, A. Martins, H. Ferreira, B. Ferreira, J. Alves, A. Dias, S. Fioravanti, D. Bertin, G. Moreno, J. Cordero, M. M. Marques, A. Grati, H. M. Chaudhary, B. Sheers, Y. Riobo, P. Letier, M. N. Jimenez, M. A. Esbri, P. Musialik, I. Badiola, R. Goncalves, A. Coelho, T. Pfister, K. Majek, M. Pelka, A. Maslowski, and R. Baptista, *Search and Rescue Robotics - From Theory to Practice*. InTech, Aug. 2017.

[11] B. Arbanas, A. Ivanovic, M. Car, M. Orsag, T. Petrovic, and S. Bogdan, "Decentralized planning and control for UAV-UGV cooperative teams," *Autonomous Robots*, vol. 42, no. 8, pp. 1601–1618, 2018.

[12] I. Noda and M. Hatayama, "Common frameworks of networking and information-sharing for advanced rescue systems," in *2004 IEEE International Conference on Robotics and Biomimetics*, pp. 245–249, 2004.

[13] J. Redmon and A. Farhadi, "Yolov3: An incremental improvement," *CoRR*, vol. abs/1804.02767, 2018.

[14] M. Nieuwenhuisen, M. Beul, R. A. Rosu, J. Quenzel, D. Pavlichenko, S. Houben, and S. Behnke, "Collaborative object picking and delivery with a team of micro aerial vehicles at MBZIRC," in *2017 European Conference on Mobile Robots (ECMR)*, IEEE, Sept. 2017.

[15] W. Hess, D. Kohler, H. Rapp, and D. Andor, "Real-time loop closure in 2D LIDAR SLAM," in *2016 IEEE International Conference on Robotics and Automation (ICRA)*, pp. 1271–1278, 2016.

[16] R. Milijas, L. Markovic, A. Ivanovic, F. Petric, and S. Bogdan, "A comparison of lidar-based slam systems for control of unmanned aerial vehicles," 2020.

[17] C. Rösmann, F. Hoffmann, and T. Bertram, "Integrated online trajectory planning and optimization in distinctive topologies," *Robotics and Autonomous Systems*, vol. 88, pp. 142–153, Feb. 2017.

[18] H. Pham and Q. Pham, "A new approach to time-optimal path parameterization based on reachability analysis," *IEEE Transactions on Robotics*, vol. 34, pp. 645–659, June 2018.

[19] D. Puljiz, M. Varga, and S. Bogdan, "Stochastic search strategies in 2d using agents with limited perception," *IFAC Proceedings Volumes*, vol. 45, no. 22, pp. 650–654, 2012.

[20] A. Batinović, J. Oršulić, T. Petrović, and S. Bogdan, "Decentralized strategy for cooperative multi-robot exploration and mapping," *IFAC-PapersOnLine*, vol. 53, no. 2, pp. 9682–9687, 2020.

[21] A. Batinovic, T. Petrovic, A. Ivanovic, F. Petric, and S. Bogdan, "A multi-resolution frontier-based planner for autonomous 3d exploration," *IEEE Robotics and Automation Letters*, vol. 6, no. 3, pp. 4528–4535, 2021.

[22] C. Zhu, R. Ding, M. Lin, and Y. Wu, "A 3D frontier-based exploration tool for mavs," in *2015 IEEE 27th International Conference on Tools with Artificial Intelligence (ICTAI)*, pp. 348–352, 2015.

[23] D. Stormont, A. Bhatt, B. Boldt, S. Skousen, and M. Berkemeier, "Building better swarms through competition: lessons learned from the aaai/robocup rescue robot competition," in *Proceedings 2003 IEEE/RSJ International Conference on Intelligent Robots and Systems (IROS 2003) (Cat. No.03CH37453)*, vol. 3, pp. 2870–2875 vol.3, 2003.

[24] L. Marconi, C. Melchiorri, M. Beetz, D. Pangercic, R. Siegwart, S. Leutenegger, R. Carloni, S. Stramigioli, H. Bruyninckx, P. Doherty, A. Kleiner, V. Lippiello, A. Finzi, B. Siciliano, A. Sala, and N. Tomatis, "The sherpa project: Smart collaboration between humans and ground-aerial robots for improving rescuing activities in alpine environments," in *2012 IEEE International Symposium on Safety, Security, and Rescue Robotics (SSRR)*, pp. 1–4, 2012.

[25] N. Miskovic, M. Bibuli, A. Birk, M. Caccia, M. Egi, K. Grammer, A. Marroni, J. Neasham, A. Pascoal, A. Vasilijevic, and Z. Vukic, "Overview of the FP7 project "CADDY - Cognitive Autonomous Diving Buddy"," in *OCEANS 2015 - Genova*, IEEE, May 2015.

[26] M. Krizmancic, B. Arbanas, T. Petrovic, F. Petric, and S. Bogdan, "Cooperative aerial-ground multi-robot system for automated construction tasks," *IEEE Robotics and Automation Letters*, vol. 5, pp. 798–805, Apr. 2020.

[27] B. Yamauchi, "A frontier-based approach for autonomous exploration,"

in *Proceedings 1997 IEEE International Symposium on Computational Intelligence in Robotics and Automation CIRA'97.*, pp. 146–151, 1997.

[28] M. A. Fischler and R. C. Bolles, "Random sample consensus: A paradigm for model fitting with applications to image analysis and automated cartography," *Commun. ACM*, vol. 24, p. 381–395, June 1981.

[29] "ERL-ER seville 2019 competition results." https://sites.google.com/catec.aero/erl-emergency-2019/scores. Accessed: 2021-07-01.

[30] A. Wegierska, K. Andrzejczak, M. Kujawinski, and G. Granosik, "Using labview and ros for planning and coordination of robot missions, the example of erl emergency robots and university rover challenge competitions," *Journal of Automation, Mobile Robotics and Intelligent Systems*, vol. 13, pp. 68–81, 09 2019.

[31] "MBZIRC competition results." https://www.mbzirc.com/winning-teams/2020/. Accessed: 2021-07-01.

[32] S. Martinez-Rozas, R. Rey, D. Alejo, D. Acedo, J. A. Cobano, A. Rodriguez-Ramos, P. Campoy, L. Merino, and F. Caballero, "Skyeye team at MBZIRC 2020: A team of aerial and ground robots for GPS-denied autonomous fire extinguishing in an urban building scenario," *CoRR*, vol. abs/2104.01834, 2021.

[33] V. Spurny, V. Pritzl, V. Walter, M. Petrlik, T. Baca, P. Stepan, D. Zaitlik, and M. Saska, "Autonomous firefighting inside buildings by an unmanned aerial vehicle," *IEEE Access*, vol. 9, pp. 15872–15890, 2021.

Chapter 5

Theory of Control Stochastic Systems with Unsolved Derivatives

Igor N. Sinitsyn

Abstract

Various types of stochastic differential systems with unsolved derivatives (SDS USD) arise in problems of analytical modeling and estimation (filtering, extrapolation, etc.) for control stochastic systems, when it is possible to neglect higher-order time derivatives. Methodological and algorithmic support of analytical modeling, filtering, and extrapolation for SDS USD is developed. The methodology is based on the reduction of SDS USD to SDS by means of linear and nonlinear regression models. Two examples that are illustrating stochastic aspects of methodology are presented. Special attention is paid to SDS USD with multiplicative (parametric) noises.

Keywords: analytical modeling, estimation (filtering, extrapolation), normal approximation method (NAM), regression (linear, nonlinear), stochastic differential systems with unsolved derivatives (SDS USD)

1. Introduction

Approximate methods of analytical modeling (MAM) of the wideband stochastic processes (StP) in stochastic differential systems with unsolved derivatives (SDS USD) based on normal approximate method (NAM), orthogonal expansions method, and quasi moment methods are developed in [1, 2]. For stochastic integrodifferential systems with unsolved derivatives reducible to SDS corresponding equations for MAM are given in [3, 4]. In [3, 4], problems of mean square (m.s.) synthesis of normal (Gaussian) estimators (filters, extrapolators, etc.) were firstly stated and solved in [1–4]. Results presented in [1–4] are valid for smooth (in m.s. sense) functions in SDS USD. For unsmooth functions in SDS USD theory of normal filtering and extrapolation is developed in [5].

Let us present an overview and generalization of [1–5] for linear and nonlinear regression models. Section 2 is developed to normal analytical modeling algorithms. Normal linear filtering and extrapolation algorithms are given in Sections 3 and 4. Linear modeling and estimation algorithms for SDS USD with multiplicated (parametric) noises are presented in Section 5. Normal nonlinear algorithms for filtering and extrapolation are described in Section 6. Section 7 contains two illustrative examples. In Section 8, main conclusions and some generalizations are given.

2. Normal modeling

Different types of SDS USD arise in problems of analytical modeling and estimator design for stochastic nonlinear dynamical systems when it is possible to neglect higher-order time derivatives [1–3].

First-order SDS USD is described by the following scalar equation:

$$\varphi = \varphi\left(t, X_t, \dot{X}_t, U_t\right) = 0, \tag{1}$$

where X_t and $\dot{X}_t$ are scalar state variable and its time derivative; U_t is noise vector StP $\left(\dim U_t = n^U\right)$; nonlinear function φ admits regression approximation [6–8].

For vector SDS USD, we have the following vector equation:

$$\overline{\varphi} = \overline{\varphi}\left(t, X_t, \overline{X}_t, U_t\right) = 0. \tag{2}$$

Here $\overline{X}_t$ being vector of derivatives till l order

$$\overline{X}_t = \left[\dot{X}_t^T \dots X_t^{(l-1)T}\right]^T; \tag{3}$$

U_t being autocorrelated noise vector defined by linear vector equation:

$$\dot{U}_t = a_{0t}^U + a_{1t} U_t + b_t^U V_t, \tag{4}$$

where $\dim X_t = n^X$; $\dim U_t = n^U$; V_t is white noise, $\dim V_t = n^V$; $\dim a_{0t}^U = n^U \times 1$; $\dim a_t^U = n^U \times n^U$; $\dim b_t^U = n^U \times n^V$. Further, we consider the Wiener white noise W_{0t} with matrix intensity $v_0 = v_0(t)$ and the mixed Wiener-Poisson white noise [9–13]:

$$V_t = \dot{W}_t, \quad W_t = W_{0t} + \int_{R_0^q} c(\rho) P^0(t, d\rho), \tag{5}$$

$$v_t = v_{0t}^W + \int_{R_0^q} c(\rho)[c(\rho)]^T v_P(t, \rho) d\rho. \tag{6}$$

Here, $\dim c(\rho) = \dim W_{0t} = n^V$; stochastic Ito integrals are taken in R_0^q (R_0^q with pricked origin).

As it is known [6–8], a deterministic model for real StP defined by $Y = \varphi(Z)$ at $Z = \left[X^T \overline{X}^T U^T\right]^T$ in (2) is given by the formula

$$\hat{y}(z) = E[Y|z], \quad \hat{y}(z) \in \Psi \tag{7}$$

at accuracy criterion

$$\varepsilon(z) = \sum_{p=1}^{n^Y} E\left[\left|\hat{y}_p - Y_p\right|^2 |z\right], p = \left[1, \dots, n^Y\right]. \tag{8}$$

Class of functions $\psi \in \Psi$ represents linear functional space satisfying the following necessary and sufficient conditions:

$$\operatorname{tr} E\left\{\left[|\hat{y}(z) - Y|\right] \psi(z)^T\right\} = 0. \tag{9}$$

For linear shifted and unshifted regression models, we have two known models:

$$\hat{y}(z) = g^B z, \quad g^B = \Gamma_{yz}\Gamma_z^{-1} \tag{10}$$

(Booton [6–8]),

$$\hat{y}(z) = a + g^K z^0, \quad g^K = K_{yz}K_z^{-1}, \quad a = E^Y - g^K E^z \tag{11}$$

(Kazakov [6–8]),
where E^z, Γ_z, K_z being first and second moments for given one-dimensional distribution.

For Eq. (2), linear regression model takes the Booton form

$$\hat{\varphi} = \hat{\varphi}_0 + k_1^{\varphi} X_t + k_2^{\varphi}\overline{X}_t + k_3^{\varphi} U_t = 0, \tag{12}$$

where $\hat{\varphi}_0, k_{1,2,3}^{\varphi}$ being regressors depending on φ and joint distribution of StP $X_t, \overline{X}_t, U_t$. After Eq. (12) differentiation till the $(l-1)$ order, we get the following set of $(l-1)$ equations:

$$\dot{\hat{\varphi}}_t = 0, \ldots, \hat{\varphi}_t^{(l-1)} = 0. \tag{13}$$

At algebraic solvability condition of linear Eqs. (12) and (13), we reduce SDS USD to SDS of the following form:

$$\dot{X}_t = A_0 + A_1 X_t + A_2 U_t, \tag{14}$$

where A_0, A_1, A_2 are expressed in terms $\hat{\varphi}_0, k_{1,2,3}^{\varphi}$ $\left(\det\left(k_2^{\varphi}\right)^{-1} \neq 0\right)$ and indirectly depends on statistical characteristics of X_t, its derivatives and noise U_t. For combined vector $\left[X_t^T U_t^T\right] = \tilde{Y}_t$ we have equation:

$$\dot{\tilde{Y}}_t = B_0 + B_1\tilde{Y}_t + B_2 V_t, \quad Y_{t0} = Y_0, \tag{15}$$

Its one and second probabilistic moments satisfy the following equations [12–14]:

$$\dot{\tilde{Y}}_t = B_0 + B_1\tilde{Y}_t + B_2 V_t, \quad Y_{t0} = Y_0, \tag{16}$$

$$\dot{E}_t^{\tilde{Y}} = B_0 + B_1 E_t^{\tilde{Y}}, \quad E_{t0}^{\tilde{Y}} = E_0^{\tilde{Y}}, \tag{17}$$

$$\dot{K}_t^{\tilde{Y}} = B_1 K_t^{\tilde{Y}} + K_t^{\tilde{Y}} B_1^T + B_2 v B_2^T, \quad \dot{K}_{t0}^{\tilde{Y}} = K_0^{\tilde{Y}}, \tag{18}$$

$$\frac{\partial K^{\tilde{Y}}(t_1, t_2)}{\partial t_2} = K^{\tilde{Y}}(t_1, t_2) B_{1t_2}^T, \quad K^{\tilde{Y}}(t_1, t_1) = K_{t_1}^{\tilde{Y}} \tag{19}$$

where $E_t^{\tilde{Y}} = E\left[\tilde{Y}_t\right], K_t^{\tilde{Y}} = E\left[\left(\tilde{Y}_t - \tilde{E}_t^Y\right)\left(\tilde{Y}_t - \tilde{E}_t^Y\right)^T\right], (t_1 > t_2)$. So, we get two proposals.

Proposal 1. *Let vector non-Gaussian SDS USD (2) satisfy conditions:*

i. *vector functions φ in Eq. (2) admit m.s. regression of linear class Ψ;*

ii. *linear Eqs. (12) and (13) are solvable regards all derivatives till $(l-1)$ order.*

Then SDS USD may be reduced to parametrized SDE. First and second moments of joint vector $\tilde{Y}_t = \left[X_t^T U_t^T\right]^T$ satisfy Eqs. (16)–(19).

Proposal 2. *For normal joint distribution $\mathcal{N} = \mathcal{N}(E_t^Y, K_t^Y)$ of vector variables in Eqs. (16)–(19) it is necessary in equations of Theorem 1 to put*

$$W_t = W_{0t},\quad E_t^{\tilde{Y}} = E_{\mathcal{N}}^{\tilde{Y}},\quad K_t^{\tilde{Y}} = E_{\mathcal{N}}^{\tilde{Y}}\left[\left(\tilde{Y}_t - E_{\mathcal{N}}^{\tilde{Y}}\right)\left(\tilde{Y}_t - E_{\mathcal{N}}^{\tilde{Y}}\right)^T\right],$$
$$K^{\tilde{Y}}(t_1,t_2) = E_{\mathcal{N}}\left[\left(\tilde{Y}_{t_1} - E_{\mathcal{N}t_1}^{\tilde{Y}}\right)\left(\tilde{Y}_{t_2} - E_{\mathcal{N}t_2}^{\tilde{Y}}\right)^T\right]. \tag{20}$$

For Eq. (2) using Kazakov form

$$\overline{\varphi} = \overline{\varphi}_0 + \overline{\varphi}^0 = 0 \tag{21}$$

where

$$\overline{\varphi}^0 = k_1^{\overline{\varphi}} X_t^0 + k_2^{\overline{\varphi}} \overline{X}_t^0 + k_3^{\overline{\varphi}} U_t^0, \tag{22}$$

we have two sets of equations for mathematical expectations and centered variables:

$$\dot{\overline{\varphi}}_0 = 0,\ \dots, \overline{\varphi}_0^{(l-1)} = 0 \tag{23}$$

$$\dot{\overline{\varphi}}_0 = 0,\ \dots, \overline{\varphi}_0^{0(l-1)} = 0. \tag{24}$$

So, we reduce SDS USD to two sets of equations for E_t^X and $X_t^0 = X_t - E_t^X$

$$\dot{E}_t^X = \overline{A}_0 + \overline{A}_1 E_t^X + \overline{A}_2 E_t^U, \tag{25}$$

$$\dot{X}_t^0 = A_1 X_t^0 + A_2 U_t^0. \tag{26}$$

For the composed vector $\overline{Y}_t^0 = \left[X_t^{0T} U_t^{0T}\right]^T$ its probabilistic one and second moments satisfy the following equations:

$$E_t^{\overline{Y}} = \overline{B}_0 + \overline{B}_1 E_t^{\overline{Y}},\quad \overline{Y}_{t0} = \overline{Y}_0, \tag{27}$$

$$\dot{K}^{\overline{Y}} = \overline{B}_1 K_t^{\overline{Y}} + K^{\overline{Y}} \overline{B}_1^T + \overline{B}_2 v \overline{B}_2^T,\quad K_{t0}^{\overline{Y}} = K_0^{\overline{Y}}, \tag{28}$$

$$\frac{\partial K^{\overline{Y}}(t_1,t_2)}{\partial t_2} = K^{\overline{Y}}(t_1,t_2)\overline{B}_{1t_2}^T,\quad K^{\overline{Y}}(t_1,t_1) = K_{t_1}^{\overline{Y}},\quad t_2 > t_1. \tag{29}$$

Here $v = v_0$ being defined by Eq. (6).

So for Kazakov regression, Eqs. (21)–(24) are the basis of Proposal 3.

The regression $E^y(z)$ and its m.s. estimator $\hat{y}(z)$ represent deterministic regression model. So to obtain a stochastic regression model, it is sufficient to represent Y in the form $Y = E^y(z) + Y'$ or $Y = \hat{y}(z) + Y''$, where Y', Y'' being some random variables. For finding a deterministic linear regression model, it is sufficient to know the mathematical expectations E^z, E^y and covariance matrices K^z, K^{yz}. In the case of a stochastic linear regression model, it is necessary to know the distribution of Y for any z or at list its regression $\hat{y}(z)$ and covariance matrix $K^y(z)$ (coinciding with the covariance matrices $K^{Y'}(z)$ or $K^{Y''}(z)$). A more general problem of the best m.s. approximation of the regression by a finite linear combination of given functions $\chi_1(z), \dots, \chi_N(z)$ is reduced to the problem of the best approximation to the regression, as any linear combination of the functions $\chi_1(z), \dots, \chi_N(z)$ represents a linear function of variables $z_1 = \chi_1(z), \dots, z_N(z) = \chi_N(z)$. Corresponding models based on m.s. optimal regression are given in [7].

In the general case, we have the following vector equation:

$$\dot{Z}_t = a^z(Z_t,t) + b^z(Z_t,t)V_t, \tag{30}$$

where V_t being defined by Eqs. (5) and (6). Functions $a^z = a^z(Z_t,t)$ and $b^z = b^z(Z_t,t)$ are composed on the basis of Eq. (2) after nonlinear regression approximation $\hat{\varphi}_t = \sum_j c_j\chi(Z_t)$ and Eq. (13).

According to normal approximation method (NAM), we have for Eq. (30) the following equations for normal modeling [9–12]:

$$\dot{E}_t^z = F_1(E_t^z, K_t^z, t), \tag{31}$$

$$\dot{K}_t^z = F_2(E_t^z, K_t^z, t), \tag{32}$$

$$\frac{\partial K_t^z(t_1,t_2)}{\partial t_2} = F_3(E_t^z, K^z(t_2), K^z(t_1,t_2), t_1, t_2). \tag{33}$$

Here

$$F_1(E_t^z, K_t^z, t) = E_N a^z(Z_t,t), \tag{34}$$

$$F_2(E_t^z, K_t^z, t) = F_{21}(E_t^z, K_t^z, t) + F_{21}(E_t^z, K_t^z, t)^T + F_{22}(E_t^z, K_t^z, t), \tag{35}$$

$$F_{21}(E_t^z, K_t^z, t) = E_N a^z(Z_t,t)(Z_t - E_t^z), \tag{36}$$

$$F_{22}(E_t^z, K_t^z, t) = E_N b^z(Z_t,t) v b^z(Z_t,t)^T, \tag{37}$$

$$\begin{aligned} &F_3\left(E_{t_2}^z, K_{t_2}^z, K^z(t_1,t_2), t\right) = K^z(t_1,t_2)\left(K_{t_2}^z\right)^{-1} F_{21}\left(E_{t_2}^z, K_{t_2}^z, t_2\right)^T, \\ &E_{t_0}^z = E^Z(t_0), \quad K_{t_0}^z = K^Z(t_0), \quad K^z(t_1,t_2) = K_{t_1}^z \end{aligned} \tag{38}$$

where E_N being symbol of normal mathematical expectation.

3. Normal linear filtering

In filtering SDS USD problems, we use two types of equations: reduced SDE USD for vector state variables X_t and equation for vector observation variables Y_t and $\dot{Y}_t \equiv Z_t$.

Consider SDS USD Eq. (2) reducible to SDE Eq. (3.9) at conditions of Theorem 1. We introduce new variables putting $X_t \equiv \tilde{Y}_t$,

$$\dot{X}_t = A_{0t} + A_{1t}X_t + A_{2t}V_{1t}. \tag{39}$$

Let the observation vector variable Y_t satisfy the following linear equations:

$$Z_t = \dot{Y}_t = B_{0t} + B_{1t}X_t + B_{2t}V_{2t}. \tag{40}$$

where V_{1t} and V_{2t} are normal white noises with matrix $v_{1t} = v_{01}$ and $v_{2t} = v_{02}$ intensities.

Equations of Kalman-Bucy filter in case of Eqs. (39) and (40) for the Gaussian white noises are as follows [12–14]:

$$\dot{\hat{X}}_t = A_0 + A_1\hat{X}_t + \beta_t\left[Z_t - \left(B_0 + B_1\hat{X}_t\right)\right]. \tag{41}$$

$$\beta_t = R_t B_{1t}^T v_{2t}^{-1}, \quad \det v_{2t} \neq 0. \tag{42}$$

$$\dot{R}_t = A_{1t} R_t + R_t A_{1t}^T + v_{1t} - \beta_t v_{2t} \beta_t^T \tag{43}$$

at corresponding initial conditions. R_t being m.s. covariance matrix error, β_t being gain coefficient. So, we have the following result.

Proposal 4. *Let:*

i. *USD are reducible to SDS according to Proposal 2 or Proposal 3;*

ii. *observations are performed according to Eq.* (40).

Then equations for m.s. normal filtering have the generalized Kalman-Bucy filter of the form (41)–(43).

4. Normal linear extrapolation

Using equations of linear m.s. extrapolation for time interval Δ [12–14] we get the following equations for the generalized Kalman–Bucy extrapolator:

$$\dot{\hat{X}}_{t+\Delta t|t} = A_1 \hat{X}_{t+\Delta t|t} \quad (\Delta > 0) \tag{44}$$

with initial condition

$$\left[\hat{X}_{t+\Delta t|t}\right]_{\Delta=0} = \hat{X}_t. \tag{45}$$

For the initial time moment t and for the final time moment $t + \Delta$ according to Eq. (44), we get

$$X_{t+\Delta t|t} = u(t+\Delta, t)X_t + \int_t^{t+\Delta} u(t+\Delta, \tau)a_0(\tau)d\tau + \int_t^{t+\Delta} u(t+\Delta, \tau)\psi(\tau)dW(\tau). \tag{46}$$

where $u(t, \tau)$ being the fundamental solution of equation $\dot{u}_t = A_{1t} u_t$ at condition $u(t, t) = I$. For conditional mathematical expectation relatively $Y_{t_0}^t$ in Eq. (46), we get m.s. estimate future state $X_{t+\Delta}$

$$\hat{X}_{t+\Delta t|t} = \mathrm{E}\left[X_{t+\Delta|t} | Y_{t_0}^t\right] = u(t+\Delta, t)\hat{X}_{t|t} + \int_t^{t+\Delta} u(t+\Delta, \tau)a_0(\tau)d\tau. \tag{47}$$

In this case, error covariance matrix $R_{t+\Delta t|t}$ satisfies the following equation:

$$\dot{R}_{t+\Delta t|t} = a_1 R_{t+\Delta|t} + R_{t+\Delta|t} a_1^T + \psi v_0 \psi^T. \tag{48}$$

At initial condition

$$\left[R_{t+\Delta|t}\right]_{\Delta=0} = R_t. \tag{49}$$

Hence, the error matrix R_t is known from Proposal 4. So, we have the following proposition.

Proposal 5. *At conditions of Proposal 4 m.s. normal extrapolation* $\hat{X}_{t+\Delta t|t}$ *is defined by Eqs. (47)–(49).*

This extrapolator presents a sequel connection m.s. filter with gain $u(t+\Delta, t)$, summator $u(t+\Delta, t)\hat{X}_{t|t}$ and integral term $\int_t^{t+\Delta} u(t+\Delta, \tau)A_{0\tau}d\tau$. The accuracy of extrapolation is estimated according to Eqs. (48) and (49).

5. Linear modeling and estimation in SDS USD with multiplicated noises

Let us consider vector Eqs. (2)–(6) for the multiplicative Gaussian noises:

$$\varphi = \varphi(\dot{X}_t, X_t, V_t) = \varphi_1(\dot{X}_t, t) + \left[\varphi_{20}(t) \sum_{h=1}^{n^X} \varphi_{2h}(t) X_h\right] V_t = 0. \quad (50)$$

Here, $\dim X_t = \dim \dot{X}_t = n^X$, $\dim \varphi = n^X$, φ_1 being nonlinear vector function of vector argument $\dot{X}_t$ admitting linear regression

$$\varphi_1(\dot{X}_t, t) \approx \varphi_{11} \dot{X}_t, \quad \varphi_{11} = \varphi_{11}\left(E_t^{\dot{X}}, K_t^{\dot{X}}, t\right). \quad (51)$$

Here, φ_{11} being matrix of regressors; V_{1t} being vector Gaussian white noise, $\dim V_t = n^V$ with matrix intensity $v = v_0(t)$. In this case, Eqs. (50) and (51) at condition $\det \varphi_{11} \neq 0$ may be resolved relatively $\dot{X}_t$

$$\dot{X}_t = B_0 + B_1 X_t + \left(B_2 + \sum_{r=1}^{n^X} B_{3r} X_{rt}\right) V_t, \quad (52)$$

where B_0, B_1, B_2, B_{3r} depend upon regressors φ_{11}. Using [9–12], we get equations for mathematical expectations. E_t^X, covariance matrix K_t^X, and matrix of covariance functions $K^X(t_1, t_2)$:

$$\dot{E}_t^X = B_0 + B_1 E_t^X, \quad E_{t0}^X = E_0^X, \quad (53)$$

$$\dot{K}_t^X = B_1 K_t^X + K_t^X B_1^T + B_2 v_0 B_2^T + \sum_{r=1}^{n^X} (B_{3r} v_0 B_2^T + B_2 v_0 B_{3r}^T)$$
$$E_{rt}^X \sum \sum_{r,s=1}^{n^X} B_{3r} v_0 B_{3s}^T (E_{rt}^X E_{st}^X + K_{rst}^X), \quad K_{t0}^X = K_0^X, \quad (54)$$

$$\frac{\partial K^X(t_1, t_2)}{\partial t_2} = K^X(t_1, t_2) B_{1t_2}^T, \quad K^X(t_1, t_1) = K_{t_1}^X, \quad t_2 > t_1. \quad (55)$$

Here $K_t^X = \left[K_{rst}^X\right]; K^X(t_1, t_2) = \left[K_{rs}^X(t_1, t_2)\right]$. So for MAM in nonstationary regimes, we have Eqs. (54) and (55) Proposal 6. In stationary case Eqs. (54) and (55) we get the following finite set of equations for E_* and K_* (Proposal 7):

$$B_0^* + B_1^* E_*^X = 0, \quad (56)$$

$$B_1^* K_*^X + K_*^X B_1^* + B_2^* v_0^* B_2^{*T} + \sum_{r=1}^{n^X} (B_{3r}^* v_0^* B_2^{*T} + B_2^* v_0^* B_{3r}^{*T}) E_*^X$$
$$+ \sum \sum_{r,s=1}^{n^X} B_{3r}^* v_0^* B_{3s}^{*T} (E_{r*}^X E_{s*}^X + K_{rs*}^X) = 0 \quad (57)$$

and ordinary differential equation for $k^X(\tau)$, $(\tau = t_2 - t_1)$:

$$\frac{dk^X(\tau)}{d\tau} = B_1^* k^X(\tau), \quad k^X(0) = K_*. \tag{58}$$

Applying linear theory Pugachev (conditionally m.s. optimal) filtering [9–12] to equations

$$\dot{X}_t = A_0 + A_1 X_t + \left(A_2 + \sum_{r=1}^{n^X} A_{3r} X_{rt}\right) V_t, \tag{59}$$

$$Z_t = \dot{Y}_t = B_0 + B_1 X_t + B_2 V_t, \tag{60}$$

We get the following normal filtering equations:

$$\dot{\hat{X}}_t = A_0 + A_1 \hat{X}_t + \beta_t \left[Z_t - \left(B_{0t} + B_1 \hat{X}_t\right)\right], \tag{61}$$

$$\beta_t = \left[R_t B_1 + \left(A_2 + \sum_{r=1}^{n^X+n^Y} A_{3r} E_r^X\right) v_0 B_2\right] \kappa_{11}^{-1}, \tag{62}$$

$$\dot{R}_t = A_1 R_t + R_t A_1^T - \left[R_t B_1^T + \left(A_2 + \sum_{r=1}^{n^X+n^Y} A_{3r} E_r^X\right) v_0 B_2\right] \kappa_{11}^{-1} \times$$
$$\times \left[B_1 + B_2 v_0 \left(A_2^T + \sum_{r=1}^{n^X+n^Y} A_{3r}^T E_r^X\right)\right] + \left(A_2 + \sum_{r=1}^{n^X+n^Y} A_{3r} E_r^X\right) v_0 \left(A_2^T + \sum_{r=1}^{n^X+n^Y} A_{3r}^T E_r^X\right)$$
$$+ \sum \sum_{r,s=1}^{n^X+n^Y} A_{3r} v_0 A_{3s}^T K_{rs}. \tag{63}$$

Here

$$\kappa_{11} = B_2 v_0 B_0^T, \quad \kappa_{22} = B_2 v_0 B_2^T. \tag{64}$$

For calculating (62) we need to find mathematical expectation E_t^Q, covariance matrix K_t^Q of combined vector $Q_t = [X_1, \dots, X_{n^X}, \ Y_1, \dots, Y_{n^Y}]^T$ and error $\tilde{X}_t$, $\tilde{X}_t = \tilde{X}_t - X_t$ covariance matrix R_t using equations

$$\dot{E}_t^Q = a^Q E_t^Q + a_0^Q, \tag{65}$$

$$\dot{K}_t^Q = a^Q K_t^Q + K_t^Q \left(a_0^Q\right)^T + c^Q v_0 \left(c^Q\right)^T + \sum_{r=1}^{n^X+n^Y} \left[c^Q v_0 \left(c_r^Q\right)^T + c_r^Q v_0 \left(c_0^Q\right)^T\right] E_r^Q +$$
$$+ \sum \sum_{r,s=1}^{n^X+n^Y} c_r^Q v_0 \left(c_s^Q\right)^T \left(E_r^Q E_s^Q + K_{rs}^Q\right), \tag{66}$$

where

$$a^Q = \begin{bmatrix} 0 & B_1 \\ 0 & A_1 \end{bmatrix}, \quad a_0^Q = \begin{bmatrix} B_0 \\ A_0 \end{bmatrix}, \quad c_r^Q = \begin{bmatrix} B_2 \\ A_{1r} \end{bmatrix} \quad \left(r = 0, 1, \dots, n^X + n^Y\right). \tag{67}$$

So, Eqs. (61)–(67) define linear Pugachev filter for SDS USD with multiplicative noises reduced to SDS (59) and (60) (Proposal 8).

At last following [9–12] let us consider linear Pugachev extrapolator for reduced SDS USD. Taking into account equations

$$\dot{X}_t = A_0 + A_1 X_t + \left(A_2 + \sum_{r=1}^{n^X} A_{3,n^Y+r} X_{rt} \right) V_1, \tag{68}$$

$$Z_t = \dot{Y}_t = B_0 + B_1 X_t + B_2 V_2 \tag{69}$$

($V_{1,2}$ being independent normal white noises with $v_{1,2}$ intensities) and the corresponding result (Section 5) we come to the following equation:

$$\dot{\hat{X}}_t = A_0(t+\Delta) + A_1(t+\Delta)\hat{X}_t + \beta_t \left[Z_t - \left(B_0 + B_1 \varepsilon_t^{-1} \hat{X}_t - B_1 \varepsilon_t^{-1} h_t \right) \right]. \tag{70}$$

Here $\varepsilon_t = u(t+\Delta, t)$, $u(s,t)$ being fundamental solution of equation $du/ds = A_1(s)$,

$$h_t = h(t) = \int_t^{t+\Delta} u(t+\Delta, \tau) A_0(\tau) d\tau. \tag{71}$$

Accuracy of linear Pugachev extrapolator (70) is performed by integration of the following equation:

$$\dot{R}_t = A_1(t+\Delta)R_t + R_t A_1(t+\Delta)^T - \beta_t (B_2 v_1 B_2^T) \beta_t^T + \left[A_2(t+\Delta) + \right.$$
$$\left. + \sum_{r=n^Y+1}^{n^X+n^Y} A_{3r}(t+\Delta) E_r(t+\Delta) \right] v_2(t+\Delta) \left[A_2(t+\Delta)^T + \sum_{r=n^Y+1}^{n^X+n^Y} A_{3r}(t+\Delta)^T E_r(t+\Delta) \right] +$$
$$+ \sum \sum_{r,s=n^Y+1}^{n^X+n^Y} A_{3r}(t+\Delta) v_2(t+\Delta) A_{3r}^T (t+\Delta)^T K_{rs}. \tag{72}$$

Equations (70)–(72) define normal linear Pugachev extrapolator for SDS USD reduced to SDS (Proposal 9).

6. Normal nonlinear filtering and extrapolation

Let us consider SDS (2) reducible to SDS and fully observable measuring system described by the following equations:

$$\dot{X}_t = a(X_t, Y_t, \alpha, t) + b(X_t, Y_t, \alpha, t) V_0, \tag{73}$$

$$Z_t = \dot{Y}_t = a_1(X_t, Y_t, t) + b_1(X_t, Y_t, t) V_0. \tag{74}$$

Here, a, a_1, b, b_1 being known functions of mentioned variable; α being vector of parameters in Eq. (73); V_0 being normal white noise with intensity matrix $v_0 = v_0(t)$.

Using the theory of normal nonlinear suboptimal filtering [10–12], we get the following equations for $\hat{X}_t$ and R_t:

$$\dot{\hat{X}}_t = f(\hat{X}_t, Y_t, R_t, t) dt + h(\hat{X}_t, Y_t, R_t, t) dt \left[dY_t - f^{(1)}(\hat{X}_t, Y_t, R_t, t) dt \right], \tag{75}$$

$$\dot{R}_t = \left\{ f^{(2)}(\hat{X}_t, Y_t, R_t, t) - h(\hat{X}_t, Y_t, R_t, t) b_1 \nu_0 b_1^T (Y_t, t) h(\hat{X}_t, Y_t, R_t, t)^T \right\} dt + \\ + \sum_{r=1}^{n_y} \rho_r(\hat{X}_t, Y_t, R_t, t) \left[dY_r - f_r^{(1)}(\hat{X}_t, Y_t, R_t, t) dt \right]. \tag{76}$$

Here

$$f(\hat{X}_t, Y_t, R_t, t) = [(2\pi)^n |R_t|]^{-1/2} \int_{-\infty}^{\infty} a(Y_t, x, t) \exp \left\{ -\left(x^T - \hat{X}_t^T\right) R_t^{-1} (x - \hat{X}_t)/2 \right\} dx, \tag{77}$$

$$f^{(1)}(\hat{X}_t, Y_t, R_t, t) = \left\{ f_r^{(1)}(\hat{X}_t, Y_t, R_t, t) \right\} \\ = \left[(2\pi)^{n_x} |R_t| \right\}^{-1/2} \int_{-\infty}^{\infty} a_1(Y_t, x, t) \exp \left\{ -\left(x^T - \hat{X}_t^T\right) R_t^{-1} (x - \hat{X}_t)/2 \right\} dx, \tag{78}$$

$$h(\hat{X}_t, Y_t, R_t, t) = \left\{ [(2\pi)^{n_x} |R_t|]^{-1/2} \int_{-\infty}^{\infty} \left[x a_1(Y_t, x, t)^T + b \nu_0 b_1^T (Y_t, x, t) \right] \times \right. \\ \left. \times \exp \left\{ -\left(x^T - \hat{X}_t^T\right) R_t^{-1} (x - \hat{X}_t)/2 \right\} dx - \hat{X}_t f^{(1)}(\hat{X}_t, Y_t, R_t, t)^T \right\} (b_1 \nu_0 b_1^T)^{-1} (Y_t, t), \tag{79}$$

$$f^{(2)}(\hat{X}_t, Y_t, R_t, t) = [(2p)^{n_x} |R_t|\}^{-1/2} \int_{-\infty}^{\infty} (x - \hat{X}_t) a(Y_t, x, t)^T \\ + a(Y_t, x, t)\left(x^T - \hat{X}_t^T\right) + b \nu_0 b_1^T (Y_t, x, t) \times \\ \times \exp \left\{ -(x^T - \hat{X}_t^T) R_t^{-1} (x - \hat{X}_t)/2 \right\} dx, \tag{80}$$

$$\rho_r(\hat{X}_t, Y_t, R_t, t) = [(2p)^{n_x} \int_{-\infty}^{\infty} \left\{ (x - \hat{X}_t)\left(x^T - \hat{X}_t^T\right) a_r(Y_t, x, t) + \right. \\ \left. + (x - \hat{X}_t) b_r(Y_t, x, t)^T \left(x^T - \hat{X}_t^T\right) + b_r(Y_t, x, t)\left(x^T - \hat{X}_t^T\right) \right\} \\ \times \exp \left\{ -\left(x^T - \hat{X}_t^T\right) R_t^{-1} (x - \hat{X}_t)/2 \right\} dx \quad (r = \overline{1, n_y}), \tag{81}$$

$$\hat{X}_0 = \mathrm{E}_N[X_0 | Y_0], \quad R_0 = \mathrm{E}_N \left[(X_0 - \hat{X}_0)\left(X_0^T - \hat{X}_0^T\right) | Y_0 \right], \tag{82}$$

where a_r being r th element of line-matrix $\left(a_1^T - \hat{a}_1^T\right)\left(b_1 \nu_0 b_1^T\right)^{-1}$; b_{kr} being element of k th line and r th column of the matrix $b_1 \nu_0 b_1^T$; b_r being the rth column of the matrix $b_1 \nu_0 b_1^T \left(b_1 \nu_0 b_1^T\right)^{-1}$, $b_r = \left[b_{1r} \dots b_{pr}\right]$ $(r = \overline{1, n_1})$.

Proposal 10. *If vector SDS USD (2) is reducible to Eqs. (73) and (74) then Eqs. (75)–(81) at conditions (82) define normal filtering algorithm. The number of equations is equal to*

$$Q_{NAM} = n_x + \frac{n_x(n_x + 1)}{2} = \frac{n_x(n_x + 3)}{2}. \tag{83}$$

Hence, if the function a_1 is linear in X_t and function b does not depend on X_t all matrices $\rho_r = 0$ and Eq. (76) does not contain $\dot{Y}_t$ (Section 3).

Analogously Section 6 we get from [12] corresponding equations of normal conditionally optimal (Pugachev) extrapolator for reduced equations

$$\dot{X}_t = a(X_t, Y_t, t) + b(X_t, t)V_1, \tag{84}$$

$$Z_t = \dot{Y}_t = a_1(X_t, Y_t, t) + b_1(X_t, Y_t, t)V_2, \tag{85}$$

where V_1 and V_2 are normal independent white noises.

7. Examples

Let us consider scalar system

$$\varphi(\dot{X}_t, X_t) \equiv \varphi_1(\dot{X}_t) + \varphi_2(X_t) + U_{1t} = 0 \tag{86}$$

$$\dot{U}_{1t} = \alpha_{10} + \alpha_{11}U_{1t} + \beta_1 V_{1t}. \tag{87}$$

Here, $X_t, \dot{X}_t$ being state variable and its time derivative; U_{1t} being scalar stochastic disturbance; V_{1t} being scalar normal white noise with intensity v_{1t}; φ_1 and φ_2 being nonlinear functions; $\alpha_{10}, \alpha_{11}, \beta_1$ being constant parameters. After regression linearization of nonlinear functions, we have

$$\varphi_1 \approx \varphi_{10} + k_{\dot{X}}^{\varphi_1}\dot{X}_t^0, \varphi_2 \approx \varphi_{20} + k_X^{\varphi_2}X_t^0. \tag{88}$$

At condition $k_{\dot{X}}^{\varphi} \neq 0$ we get from (86) and (88) equations for mathematical expectation $m_t^X = EX_t$ and centered $X_t^0 = X_t - m_t^X$:

$$\varphi_{10} + \varphi_{20} + m_{1t}^U = 0, \tag{89}$$

$$\dot{X}_t^0 = a_t X_t^0 + b_t U_{1t}^0, \tag{90}$$

where

$$\varphi_{10} = \varphi_{10}\left(m_t^{\dot{X}}, D_t^{\dot{X}}\right), \quad \varphi_{20} = \varphi_{20}\left(m_t^X, D_t^X\right), \tag{91}$$

$$a_t = a_t\left(m_t^X, m_t^{\dot{X}}, D_t^X, D_t^{\dot{X}}, D_t^{U_1}, D_t^{XU_1}\right) = -k_X^{\varphi_2}\left(k_{\dot{X}}^{\varphi_1}\right)^{-1}, \quad b_t = -\left(k_{\dot{X}}^{\varphi_1}\right)^{-1}. \tag{92}$$

Equations (87) and (90), for $U_{1t}^0 = U_{1t} = m_t^{U_1}$ $\left(m_t^{U_1} = EU_{1t}\right)$ and $\overline{X}_t = [X_t \;\; U_{1t}]^T$ may be presented in vector form

$$\dot{m}_t^{\overline{X}} = A_{0t} + A_t m_t^{\overline{X}}, \tag{93}$$

$$\dot{\overline{X}}_t^0 = A_t \overline{X}_t^0 + B_t V_{1t}, \tag{94}$$

$$A_t = \begin{bmatrix} a_t & b_t \\ 0 & \alpha_1 \end{bmatrix}, \quad B_t = \begin{bmatrix} 0 \\ \beta_1 \end{bmatrix}. \tag{95}$$

Covariance matrix

$$K_t^{\overline{X}} = \begin{bmatrix} D_t^X & K^{XU_1} \\ K_t^{\dot{X}U_1} & D_t^{U_1} \end{bmatrix} \tag{96}$$

and matrix of covariance functions

$$K^{\overline{X}}(t_1,t_2) = \begin{bmatrix} K_{11}^{\overline{X}}(t_1,t_2) & K_{12}^{\overline{X}}(t_1,t_2) \\ K_{21}^{\overline{X}}(t_1,t_2) & K_{22}^{\overline{X}}(t_1,t_2) \end{bmatrix} \tag{97}$$

satisfy to linear equations for correlation theory (Section 3)

$$\dot{K}_t^{\overline{X}} = A_t K_t^{\overline{X}} + K_t^{\overline{X}} A_t^T + B_t \nu_{1t} B_t^T, \quad K_{t_0}^{\overline{X}} = K_0^{\overline{X}}, \tag{98}$$

$$\frac{\partial K^{\overline{X}}(t_1,t_2)}{\partial t_2} = K^{\overline{X}}(t_1,t_2) A_{t_2}^T, \quad K^{\overline{X}}(t_1,t_1) = K_{t_1}^{\overline{X}}. \tag{99}$$

Vector Eq. (98) is equal to the following scalar equations:

$$\dot{D}_t^X = 2\left(a_t D_t^X + b_t K_t^{XU_1}\right), \quad \dot{D}_t^{U_1} = 2\,\alpha_1 D_t^{U_1} + \beta_1^2 \nu_{1t}, \quad \dot{K}_t^{XU_1} = a_t K_t^{XU_1} + b_t D_t^{U_1} + \alpha_1 K_t^{XU_1}. \tag{100}$$

From Eq. (90) we calculate variance

$$D_t^{\dot{X}} = a_t^2 D_t^X + b_t^2 D_t^{U_1} + 2 a_t b_t K_t^{XU_1}. \tag{101}$$

Thus, for MAM algorithm we use Eqs. (89), (91), (98)–(101).
Let system (86) and (87) is observable so that

$$Z_t = \dot{Y}_t = X_t + V_{2t}. \tag{102}$$

Then for Kalman-Bucy filter equations (Proposal 4), we have

$$\dot{\hat{X}}_t = A_t \hat{X}_t + \beta_t\left(Z_t - \hat{X}_t\right), \quad \beta_t = R_t \nu_{2t}^{-1} \quad (\det \nu_{2t} \neq 0), \quad \dot{R}_t = 2 A_t R_t + \nu_1 - \nu_2 \beta_t^2. \tag{103}$$

For Kalman-Bucy extrapolator equations are defined by Proposal 5 at $u(t,\tau) = e^{-a(t-\tau)}$.

In **Table 1**, the coefficients of statistical linearization of for typical nonlinear function are given.

Let us consider normal scalar system

$$F \equiv \Phi\left(\dot{X}_t\right) + a_t X_t + u_t = 0. \tag{104}$$

Here random function admits Pugachev normalization

$$\Phi\left(\dot{X}_t\right) \approx \Phi_0 + k^{\Phi} \dot{X}_t^0 + \Delta\Phi_t^0, \tag{105}$$

where $\Delta\Phi_t^0$ being normal StP satisfying equation of forming filter

$$\Delta\dot{\Phi}_t^0 = a_t^{\Delta\Phi} \Delta\Phi_t^0 + b_t^{\Delta\Phi} V_t. \tag{106}$$

Note that functions Φ_t^0 and k^{Φ} depend on $E_t^{\dot{\Phi}}$ and $D_t^{\dot{\Phi}}$. Equations (104) and (105) are decomposing on two equations. First equation at condition $k^{\Phi} \neq 0$ is as follows:

$$\Phi_0 + a_t E_t^X + u_t = 0, \quad \Phi_0 = k^{\Phi_0} E_t^{\dot{X}}. \tag{107}$$

φ	φ_0
$\dot{Y}^3$	$m(m^2+3D)$
$\sin\omega\dot{Y}$	$\exp\left(-\frac{\omega^2 D}{2}\right)\sin\omega m$
$\cos\omega\dot{Y}$	$\exp\left(-\frac{\omega^2 D}{2}\right)\cos\omega m$
$\dot{Y}\exp(\alpha\dot{Y})$	$(m+\alpha D)\exp\left(\alpha m+\frac{\alpha^2 D}{2}\right)$
$\dot{Y}\sin\omega\dot{Y}$	$(m\sin\omega m-\omega D\cos\omega m)\exp\left(-\frac{\omega^2 D}{2}\right)$
$\dot{Y}\cos\omega\dot{Y}$	$(m\cos\omega m-\omega D\sin\omega m)\exp\left(-\frac{\omega^2 D}{2}\right)$
$\operatorname{sgn}\dot{Y}$	$2\Phi\left(\frac{m}{\sqrt{D}}\right)$
$\dot{Y}^2\operatorname{sgn}\dot{Y}$	$2D\left\{\left(\frac{m^2}{D}+1\right)\Phi\left(\frac{m}{\sqrt{D}}\right)+\frac{1}{\sqrt{2\pi D}}\exp\left(-\frac{m^2}{2D}\right)\right\}\quad(m=m_{\dot{Y}},\quad D=D_{\dot{Y}})$
$\begin{cases}\frac{l}{D}\dot{Y}, & \lvert\dot{Y}\rvert\le d;\\ l, & \dot{Y}<d;\\ -l & \dot{Y}<-d\end{cases}$	$l\left\{(1+m_1)\Phi\left(\frac{1+m_1}{\sigma_1}\right)-(1-m_1)\Phi\left(\frac{1-m_1}{\sigma_1}\right)+\right.$ $\left.+\frac{\sigma_1}{\sqrt{2\pi}}\left[\exp\left\{-\frac{1}{2}\left(\frac{1+m_1}{\sigma_1}\right)^2\right\}-\exp\left\{-\frac{1}{2}\left(\frac{1-m_1}{\sigma_1}\right)^2\right\}\right]\right\}$
$\begin{cases}\gamma(\dot{Y}+d), & \lvert\dot{Y}\rvert<-d;\\ 0, & \lvert\dot{Y}\rvert\le d;\\ \gamma(\dot{Y}-d) & \dot{Y}>d\end{cases}$	$\gamma\left\{1-\frac{1}{m_1}\left[(1+m_1)\Phi\left(\frac{1+m_1}{\sigma_1}\right)-(1-m_1)\Phi\left(\frac{1-m_1}{\sigma_1}\right)\right]+\right.$ $\left.+\frac{\sigma_1}{m_1\sqrt{2\pi}}\left[\exp\left\{-\frac{1}{2}\left(\frac{1-m_1}{\sigma_1}\right)^2\right\}-\exp\left\{-\frac{1}{2}\left(\frac{1+m_1}{\sigma_1}\right)^2\right\}\right]\right\}$
$\begin{cases}-l, & \dot{Y}<-d;\\ 0, & \lvert\dot{Y}\rvert\le d;\\ l & \dot{Y}>d\end{cases}$	$l\left[\Phi\left(\frac{1+m_1}{\sigma_1}\right)-\Phi\left(\frac{1-m_1}{\sigma_1}\right)\right]$
$\dot{Y}_1\dot{Y}_2$	$m_1m_2+K_{12}$
$\dot{Y}_1^2\dot{Y}_2$	$(m_1^2+K_{11})m_2+2m_1K_{12}$
$\sin(\omega_1\dot{Y}_1+\omega_2\dot{Y}_2)$	$\exp\left[\frac{\omega_1^2K_{11}+2\omega_1\omega_2K_{12}+\omega_2^2K_{22}}{2}\right]\sin(\omega_1m_1+\omega_2m_2)$
$\operatorname{sgn}(\dot{Y}_1+\dot{Y}_2)$	$2\Phi(\zeta_{1,2}),\zeta_{1,2}=\frac{m_1+m_2}{\sqrt{D}},\quad D=K_{11}+2K_{12}+K_{22}$

Table 1.
Coefficients of statistical linearization for typical nonlinear functions [12–14].

Second equation at condition $k^\Phi\neq 0$ is as follows: $k^\Phi\dot{X}_t^0+\Delta\Phi_t^0+a_tX_t^0=0$ may be presented as

$$\dot{X}_t^0=a_t\left(k^\Phi\right)^{-1}X_t^0-\left(k^\Phi\right)^{-1}\Delta\Phi_t^0. \tag{108}$$

Equations (106) and (108) for $Z_t^0=\left[X_t^0\ \ \Delta\Phi_t^0\right]^T$ leads to the following vector equation for covariance matrix

$$\dot{K}_t^Z=AK_t^Z+K_t^ZA^T+B\nu^VB^T, \tag{109}$$

where $A=\begin{bmatrix}-a_t\left(k^\Phi\right)^{-1} & -\left(k^\Phi\right)^{-1}\\ 0 & a^{\Delta\Phi}\end{bmatrix},\quad B=\begin{bmatrix}0\\ b^{\Delta\Phi}\end{bmatrix}$. Eqs. (107) and (109) give the following final relations:

$$E_t^{\dot{X}}=\left(a_tE_t^X+u_t\right)\left(k^{\Phi_0}\right)^{-1}, \tag{110}$$

$$
\begin{aligned}
D_t^{\dot{X}} &= a_t^2 \left(k^{\Phi}\right)^{-2} D_t^X + \left(k^{\Phi}\right)^{-2} D_t^{\Delta\Phi} - 2a_t \left(k^{\Phi}\right)^{-2} K_t^{X\Delta\Phi}, \\
\dot{D}_t^X &= -2\left(a_t D_t^X + K_t^{X\Delta\Phi}\right), \\
\dot{D}_t^{\Delta\Phi} &= 2a_t^{\Delta\Phi} D_t^{\Delta\Phi} + \left(b_t^{\Delta\Phi}\right)^2 \nu^V, \\
\dot{K}_t^{X\Delta\Phi} &= a_t^{\Delta\Phi} K_t^{\Delta\Phi} - \left(a_t K_t^{X\Delta\Phi} + D_t^{\Delta\Phi}\right)\left(k^{\Phi}\right)^{-1}.
\end{aligned}
\tag{111}
$$

8. Conclusion

Models of various types of SDS USD arise in problems of analytical modeling and estimation (filtering, extrapolation, etc.) for control stochastic systems, when it is possible to neglect higher-order time derivatives. Linear and nonlinear methodological and algorithmic support of analytical modeling, filtering, and extrapolation for SDS USD is developed. The methodology is based on the reduction of SDS USD to SDS by means of linear and nonlinear regression models. Special attention is paid to SDS USD with multiplicative (parametric) noises. Examples illustrating methodology are presented. The described results may be generalized for systems with stochastically unsolved derivatives and stochastic integrodifferential systems reducible to the differential.

Acknowledgements

The author is grateful to experts for their appropriate and constructive suggestions to improve this template. Research is supported by the Russian Academy of Sciences (Project-AAAA-A19-119001990037-5). Also, the author is much obliged to Mrs. Irina Sinitsyna and Mrs. Helen Fedotova for translation and manuscript preparation.

Author details

Igor N. Sinitsyn[1,2]

1 Federal Research Center “Computer Science and Control” of Russian Academy of Sciences, Moscow, Russia

2 Moscow Aviation Institute (Technical University), Moscow, Russia

*Address all correspondence to: sinitsin@dol.ru

References

[1] Sinitsyn IN. Analytical modeling of wide band processes in stochastic systems with unsolved derivatives. Informatics and its Applications. 2017; **11**(1):2-12. (in Russian)

[2] Sinitsyn IN. Parametric analytical modeling of processes in stochastic systems with unsolved derivatives. Systems and Means of Informatics. 2017;**27**(1):21-45. (in Russian)

[3] Sinitsyn IN. Normal suboptimal filters for stochastic systems with unsolved derivatives. Informatics and its Applications. 2021;**15**(1):3-10. (in Russian)

[4] Sinitsyn IN. Analytical modeling and filtering in integrodifferential systems with unsolved derivatives. Systems and Means of Informatics. 2021;**31**(1):37-56. (in Russian)

[5] Sinitsyn IN. Analytical modeling and estimation of normal processes defined by stochastic differential equations with unsolved derivatives. Mathematics and Statistics Research. 2021. (in print)

[6] Pugachev VS. Theory of Random Functions and its Application to Control Problems. Pergamon Press; 1965. p. 833

[7] Pugachev VS. Probability Theory and Mathematical Statistics for Engineers. Pergamon Press; 1984. p. 450

[8] Pugachev VS, Sinitsyn IN. Lectures on Functional Analysis and Applications. Singapore: World Scientific; 1999. p. 730

[9] Pugachev VS, Sinitsyn IN. Stochastic Differential Systems. Analysis and Filtering. Chichester: John Wiley & Sons; 1987. p. 549

[10] Pugachev VS, Sinitsyn IN. Theory of Stochastic Systems. 2nd ed. Moscow: TORUS Press; 2001. p. 1000. (in Russian)

[11] Pugachev VS, Sinitsyn IN. Stochastic Systems. Theory and Applications. Singapore: World Scientific; 2001. p. 908

[12] Sinitsyn IN. Kalman and Pugachev Filters. 2nd ed. Logos: Moscow; 2007. p. 772. (in Russian)

[13] Socha L. Linearization Methods for Stochastic Dynamic Systems, Lect Notes Phys. 730. Springer; 2008. p. 383

[14] Sinitsyn IN. Normalization of systems with stochastically unsolved derivatives. Informatics and its Applications. 2021. (in print, in Russian)